Building Quantities: worked examples

D1376101

Building Quantities: worked examples

R. J. Wheeler ARICS
A. V. Clark ARICS

Newnes
An imprint of Butterworth-Heinemann Ltd
Linacre House, Jordan Hill, Oxford OX2 8DP

PART OF REED INTERNATIONAL BOOKS

OXFORD   LONDON   BOSTON
MUNICH   NEW DELHI   SINGAPORE   SYDNEY
TOKYO   TORONTO   WELLINGTON

First published 1992

British Library Cataloguing in Publication Data
Wheeler, R. J.
   Building Quantities: worked examples
   I. Title   II. Clark, A. V.
   692

ISBN 0 7506 0280 5

Typeset by Vision Typesetting, Manchester
Printed and bound in Great Britain by
Thomson Litho Limited, East Kilbride, Scotland

Contents

Preface

The purpose of this book is to provide a practical guide comprising worked examples, rules of measurement and experience of measurement of building quantities based on SMM7. Although it has been designed principally for students studying for degrees, diplomas and certificates in building in general and surveying in particular, e.g. BTEC, RICS, CIOB, IAAS, it has also been aimed at practising surveyors/builders in the construction industry.

The authors have tried to set out principles of measurement, show the logical sequence of measurement and illustrate the correct mensurational techniques in an effort to satisfy the need for basic quality assurance, so that the reader will be able to apply and interpret relevant sections of SMM7 to the component parts of the work with reference to the drawings and specification.

Any book of reference such as the Standard Method of Measurement benefits from guidance and explanation, and the following worked examples of measurement based on SMM7 are the authors' interpretations – they are *a* model answer; they are not necessarily *the* model answer. Even within a standard method of measurement there exist alternatives, different interpretations and opinions and where the authors are aware of a different (though still acceptable) approach, this has been noted and comment made.

It is the authors' belief that if measurers can master the general principles and follow a guided systematic logical approach for one answer, then they can interpret and apply that knowledge within the context of the problem in hand and the principles of any system of measurement.

By following each of the flow charts produced for the examples measured, measurement can be carried out and assumptions built up from SMM7.

The charts show:

1. a suggested order of measurement
2. the relevant SMM7 reference and unit of measurement
3. the various measurements to give the quantity of the required units
4. how and from where the various measurements have been obtained

Where the masculine form has been used in this book this is purely an abstract form for convenience.

Acknowledgements

The authors would like to thank the Standing Joint Committee for the Standard Method of Measurement of Building Works for their permission to quote from the Standard Method of Measurement of Building Works Seventh Edition.

We would also like to thank Sally Walton and Robinson Kenning & Gallagher, Lombard Business Park, Croydon for the CAD artwork and a special word of thanks to Roy Pettitt FRICS, D. Baccarini ARICS, our wives, colleagues, friends and Bridget for their help and continued patience.

Note

1

Introduction

Why do we measure buildings? Perhaps this is a strange question to ask in a book which devotes the next 275 pages explaining the various techniques and procedures of the measurement of building quantities, but it is a valid question none the less.

Traditionally, the major role of professional quantity surveyors was to measure and prepare bills of quantities, and although emphasis on this aspect of a quantity surveyor's work has diminished over recent years it is still an important role and it is on this role that this book concentrates.

Throughout the construction of a building project, from inception to completion, different measurements may be required according to the stage of the project, the purpose of the measurements and the accuracy required.

Design stage

At an early stage of a scheme, the client normally requires an approximate estimate of the cost of the proposed building works in order to establish the economic viability of the project. This will involve the measurement of

1. the floor area (square metres)
2. the volume (cubic metres)
3. the approximate quantities, or
4. the various building elements (floors, walls, etc.)

to provide a cost plan/analysis for the scheme. As the working drawings are progressed, they will be checked to ensure that any changes compared with the information used in the preparation of the preliminary estimate are picked up.

Tender stage

From working drawings the quantity surveyor will take measurements in order to produce a bill of quantities which sets out the quantities and quality of all items required to construct the work in accepted units of measurement in a logical and recognized manner – conventionally in accordance with a Standard Method of Measurement.

Basically, the primary purpose of a bill of quantities is to obtain competitive tenders by enabling the building contractors tendering for a contract to do so on a uniform basis. Providing the same information avoids the need for each tenderer to measure the quantities himself before giving an estimate which will incorporate labour, materials and plant costs, plus overheads, waste and profit. On the advice of his consultant the client will accept the most favourable tender which may not necessarily be the lowest.

The priced bill of quantities becomes a contract document and provides a detailed breakdown of the contract sum which, when used for the valuation of work in progress and adjustment of the measurement and value of a varied work, allows for financial management of the project. In addition, as it obviously provides information on building costs, it will be analysed to provide information for the cost planning of future projects.

As the bill of quantities provides an itemized list of the component parts of the building, the contractor may use it to programme the project and calculate his resource requirements, i.e. to assess labour and plant requirements, and to order materials, although there is no contractual guarantee as to its accuracy for these purposes.

Construction stage

In the United Kingdom, most building contracts allow the architect/supervising officer to vary the work measured contained in the bills of quantities. The method of measuring these variations will generally be the same as that used for the production of the bills of quantities, although some of the measurements may be actual, i.e. physically recorded on site rather than from drawings. However, in some instances, it may not be possible to measure the work conventionally and, accordingly, such work may be recorded and valued on a day-work basis.

Interim valuations that will determine the amounts which should be paid to the contractor are based on an assessment or physical measurement of work executed during the progress of the contract.

Bill preparation

There are several methods of preparing bills of quantities.

Traditional method

For convenience this may be considered as having three distinct stages:

1. **Taking-off**   This involves the measurement of dimensions and the composition of clear and concise descriptions sufficiently detailed to allow the accurate pricing of the works.

2. **Working-up**   The process commonly known as 'working-up' concerns the collation, collection and calculation of the measured information in a recognized bill order.

3. **Editing**   As the draft bill is written direct from the abstract it is the last opportunity for the editor, prior to printing, to make checks and adjustments.

The traditional method of preparing bills of quantities is regarded as laborious and refinements and alternative processes have been introduced generally as technology has progressed.

Billing direct

Basically, this method involves the transfer of measured items from the dimension sheets to the bill without the necessity for the intermediary process known as abstracting. These items are taken off separately in the order in which they will be billed, thereby reducing time of preparation and, consequently, lowering costs. However, this system is only really suitable when the work is relatively straightforward and the number of items limited. The measurement of drainage is an area of work where the adoption of this method may be used to advantage.

Cut and shuffle method

This is a system of slip sorting the dimensions and descriptions into bill order as a way of reducing the bill preparation time, and rationalizing the traditional method. There is not one standard format although, basically, the process involves one copy of the dimensions and descriptions being prepared in bill order and another copy (usually the original) being kept in taking-off order, namely:

1. Measurement is carried out on specially printed A4 sheets divided vertically into three or possibly four sections. The taking-off is carried out conventionally but with only one description and any number of dimensions on each slip, which, for ease of reference, must be numbered. As sections of the measurement are completed, waste calculations are checked, dimensions are squared and repeat dimensions calculated. Unless the paper is self-carbonating, a photocopy of each completed dimension sheet will be required, which can then be cut to form slips.

2. The slips are sorted into the various trades or sections of the bill, placed into the correct bill order of sequence, and then reduced to a collection of quantities on what are termed 'parent' or 'master' primary slips. It should be noted that each 'parent' or 'master' slip will have as many 'child' or 'slave' auxiliary slips as are necessary, but that bill items are made only from 'parent' or 'master' slips. Slips should be numbered and a note taken of the numbers of child and waste slips. As a check the total of these should equate with the original number of slips. Accordingly on no account should nilled, waste or blank slips be discarded.

3. The primary item slips with full descriptions and quantities without abbreviations are edited, headings and sub-headings are inserted and the bill of quantities is then typed direct from the slips.

This system avoids the process of abstracting and billing and cuts down on the repetition of descriptions during measurement.

Computerized methods

Increasingly computers are being used for the preparation of bills of quantities and other quantity surveying functions as the cost of equipment and software programs, falls and technology begins to meet the real needs of the profession.

Computerized bill production eliminates the need for reducing, abstracting and billing processes although obviously the accuracy of the output will be determined by the accuracy of the input data. A standard library of descriptions should be based on the SMM7 reference codes with provision for non-standard or rogue items. Computers may be programmed to print out alternative bill formats from the original set of dimensions. For example, whilst a trade order bill is useful for tendering purposes, an elemental bill is more useful for cost planning and financial management and an operational bill would be more useful for site management where a method statement was of importance.

With the advances being made in the development and design of systems there is little doubt that the use of computers will become more widespread.

Requirements for accuracy

'The Standard Method of Measurement provides a uniform basis for measuring building works and embodies the essentials of good practice'. Therefore, whatever the method adopted for bill preparation, the requirements for accuracy will be the same. For example:

1. 'Work shall be measured net as fixed...' (General Rule 3.1).

2. 'Dimensions used in calculating quantities shall be taken to the nearest 10 mm...' (General Rule 3.2). Accordingly, all dimensions should be recorded to two decimal places and, as a consequence, waste calculations or side-casts should be calculated to three decimal places.

3. 'Quantities should be billed to the nearest whole unit except that any quantity less than one unit shall be given as one unit, and quantities measured in tonnes shall be given to two places of decimals' (General Rule 3.3).

4. 'Openings or wants which are at the boundaries of measured areas shall always be the subject of deduction, irrespective of size' (General Rule 3.4).

5. 'Items can be measured within a range of stated limits, e.g. plain vertical formwork to edge of suspended *in situ* concrete slab 200 mm high would be categorized and described as equal to or not exceeding 250 mm in height' (Rule E20.3.1.2).

6. 'Minimum deductions for voids, e.g. formwork, are not deducted from slabs and landings having voids less than or equal to 5 square metres' (Rule E20.M4).

7. Averaging or an interpolation of levels may be required for the measurement of excavation.

Due to problems with reprographics and the inherent inaccuracy of the process, scaling drawings should be avoided whenever possible. However, where dimensions are stated on drawings, there is no excuse for measurements being anything other than correct. A bill of quantities should be as accurate a reflection of the drawings and information available at the time of measurement as possible.

Content of a bill of quantities

In addition to the measured work section, the bill of quantities includes sections for Preliminary Items, Preamble Items, Prime Costs and Provisional Sums and also usually Dayworks.

Preliminaries

Section A of SMM7 provides guidance as to the items which should be considered for inclusion in this section of a bill of quantities. The preliminaries bill sets out the practical and contractual conditions of the project, i.e. location, site description, nature and the extent of the work, the type of contract and generally all factors that may affect the execution of the works, etc, together with project overheads, i.e. items that cannot easily be attributed to any particular measured work section, such as supervision, scaffolding, protection, services, facilities, security, health and welfare, etc.

Preambles

The Preamble clauses of a bill of quantities specifies for each work section:

1. quality and description of materials
2. standards of workmanship
3. other relevant information, such as testing, measurement notes, etc.

The principal advantage of this procedure is that bill descriptions can be reduced in length and repetition of information is avoided.

Prime cost and provisional sums

Prime cost ('PC') sums

A PC sum is an approximate sum of money to be included by the tenderer for works and services to be carried out by a nominated sub-contractor, usually of a specialist nature, e.g. lift installation, or for materials and goods to be obtained from a nominated supplier. SMM Sections A51 and A52 deal with the provision of these aspects of the works.

Provisional sums

Provisional sums are used to cover the cost of works, the exact nature of which is not known at the time of tendering. However, SMM General Rule 10 states that 'Where work cannot be described and given in items in accordance with these Rules it shall be given as a Provisional Sum and identified as for either defined or undefined work as appropriate'. The difference is that where work can be reasonably well described and termed as defined work the contractor will have been expected to have 'made due allowance in programming, planning and pricing preliminaries', whereas if the work is undefined it is deemed that no such allowances have been made and, therefore, the implication is that an adjustment to the final account and contract period may be justifiable.

    SMM Section A53 covers works to be carried out by local authorities and statutory undertakings, and it should be noted that work by statutory authorities includes work by public companies, e.g. a water company responsible for statutory work when executing their statutory duty, e.g. mains service supply.

    Dayworks are considered with regard to SMM Section A55 and should be made in respect of work which cannot be measured and valued by other means. Essentially, the contractor is paid on the basis of the prime cost of labour, materials and plant with an agreed percentage addition for overheads and profit generally calculated in accordance with the 'definition of prime cost of daywork carried out under a building contract' as published by the Royal Institution of Chartered Surveyors and the National Federation of Building Trades Employers.

Standard Method of Measurement (SMM)

To avoid the confusion that arose when individual quantity surveyors had their own method of measurement, an attempt at standardization was made in 1922 with the publication of the first edition of the Standard Method of Measurement of Building Works issued by the Royal

Institution of Chartered Surveyors and the National Federation of Building Trades Employers (now known as the Building Employers Confederation).

The latest edition (seventh) of this document was published in 1988 and it is to this edition, together with the accompanying Code of Procedure for the measurement of building works, that this book relates. In contrast with the traditional prose style of previous editions, SMM7 adopts a tabular format and a simplification of the rules of measurement has brought about a reduction of items in bills of quantities.

The SMM provides a uniform basis for measuring building works and embodies the essentials of good practice. However, it should be noted that more detailed information than is required by the rules shall be given where necessary in order to define the precise nature and extent of the required work.

It has been said that 'rules are for the guidance of wise men and for the observance of fools'. Sometimes in practice it may be more appropriate to depart from the rules in the SMM, in which case this must be expressly stated in the description of the item. However, generally it is stressed that the principles of measurement as laid down in the SMM should be followed as, obviously, the more deviations that take place then the less uniform a document is produced, thereby creating possible uncertainty. It is recommended that the SMM should be adhered to in the preparation of bills of quantities, the measurement of executed works and, especially, for examination purposes.

Need for change

It is acknowledged that there is a tendency among practitioners to complain about the necessity for changes in the rules of measurement. However, these complaints (of which we are only too aware) generally arise because people are conservative by nature and tend to resist change. It must be recognized that the SMM is not a static document but is an evolving one which tries to respond to the changing nature of the building industry. Having said that, it is the authors' belief that future amendments would be better received if they were confined to revisions of sections of work rather than wholesale change; this would be more easily digestible. Nevertheless there is a real need for change which is basically attributable to two factors, advances in technology and evolution.

Advances in technology

Given the continual development of building methods and materials, many changes have taken place since the inception of the first edition of the SMM, not the least being the move away from the more labour-intensive craft processes. For example, in the past only locally available materials tended to be used. Today a wide range of new materials is available which, together with new methods of construction, e.g. increased use of machinery, combined with the use of more off-site fabrication such as trussed rafters, and revised legislation such as insulation requirements, have led to a revolution in the construction industry and to the eclipse of outmoded building crafts, such as thatching. Moreover, the ever increasing sophistication of user requirements with regard to services, such as central heating and air conditioning, has brought about significant changes in the provision of what was originally only conceived as basic services.

Evolution

The requirements of the rules of measurement change as the building industry evolves. For example, the introduction of metrication, the widespread adoption of standard phraseology, the pressure for simplification (it used to be said that 90% of the value of quantities was contained in only 10% of the items), the use of computerized methods to produce bills of quantities, the greater demand upon supervision and management systems as a result of changes in working practices and, perhaps the greatest change of all, the lowering of trade restrictions within the European Economic Community in 1992, have had or will have a profound effect on the construction industry.

Quality control (QC)

The principal aim of QC is to establish confidence in an organization's capability by consistently achieving stated objectives. Much has been said or written concerning QC; basically, it is about making every individual accountable and responsible for their own efforts and this can be achieved by the use of simple procedures based on sound and practical methods.

In the context of bills of quantities, under most forms of building contract any error is not at the risk of the tenderer but will be deemed a variation and, as such, liable for adjustment in the final account. Therefore, there is an obligation to ensure that the bills of quantities are as accurate as possible by applying checks on both the process and procedure in an effort to avoid or, if necessary, correct errors at the appropriate stage.

Checklist

1. Create a drawing register for each project to ensure that only the latest revised drawings are used. Date stamp and enter all drawings in the register, clearly marking all previous drawings 'superseded', and stamp the drawings used for taking-off as 'used for bill of quants'. A complete list of these drawings should be incorporated in the preliminaries and should form part of the tender documents.

2. Check that there is sufficient information for taking-off purposes; if not, make use of query sheets which should be sent to the supervising officer for reply with a copy kept on file. It is also worthwhile to check that the correct drawings and the correct number of drawings, together with the correct information, have been issued.

3. Scales on drawings and the sum of internal and external dimensions should be arithemetically checked – there is a story about a villa in the United States being designed by an Italian architect in metric but, inadvertently, being built to what were assumed to be imperial measurements!

4. To enable the structure to be accurately measured check the various plans, elevations and sections for any inconsistencies, and ensure that the sections drawn are relevant.

5. Any separate schedules and/or specification notes should be compared with annotations on the drawings and especially with regard to references and the number of items, e.g. total number of doors and windows.

6. Check the general information to be included in the preliminaries section, e.g. type of contract, special constraints or limitations, access, etc.

7. The name of the project must appear on every sheet, which should also be numbered and usually dated.

8. Offer the taking-off in a clear, uncramped presentation and write the dimensions so as to avoid introducing errors.

    (a) A taking-off list is essential with items being marked through once they have been measured. This process provides an ordered logical framework, helps to concentrate the mind and helps to prevent omissions. It should be as long or as short as necessary but it must be adequate.

    (b) Annotate everything for clarity and make side notes for ease of reference and, as stated previously, always use figured dimensions in preference to scaling. Where drawings are not fully dimensioned, check that any scaled dimensions do not contradict the overall dimensions.

    (c) All sundry calculations should be calculated in waste so that the arithmetic can be checked and others can follow the taking-off. Mental calculations cannot be checked.

    (d) Dimensions should never be erased, obliterated or altered. To correct dimensions neatly cross them out, write 'nil' alongside, taking care to identify clearly the extent of the correction, and rewrite correctly. As it is often useful to be able to read the original figures, the use of 'liquid paper' correcting fluid is to be resisted at all costs.

    (e) All figures must be clearly written and in the correct column.

    (f) To avoid repetition of dimensions, use is made of the timesing column whereby they can be multiplied. Great care must be taken when 'timesing' or 'dotting on' so as to prevent squaring errors, particularly when dealing with fractions. Accordingly, when writing fractions in the timesing column a horizontal line, rather than a diagonal slash, must be used to avoid any confusion with timesing.

    (g) When an item or dimension is to be deducted it must be prefaced with the word 'Deduct' (more commonly in its abbreviated form, 'Ddt') and, although not strictly essential, subsequent deductions are marked similarly. However, the first item or dimension immediately following the deduction should be prefaced with the word 'Add'; when a number of items, some of which are to be deducted, use the same dimensions by virtue of the ampersand, every item must be prefaced with the word 'Deduct' or 'Add'.

    (h) Do not use 'ditto' at the head of a sheet or the head of a new page.

    (i) Centre lines used for a number of related items should be checked carefully.

    (j) Rough arithmetical checks should be made on the overall quantities which are interrelated, e.g. total excavation = backfill and disposal; area of decoration = area of plaster; ceiling = floor.

    (k) Read dimensions through for any obvious errors before offering them up for squaring and the person who squares the dimensions should first check waste calculations to ensure that in the transfer of totals to the dimension column no errors have occurred. If an error is found it should be referred to the taker-off prior to adjustment. Once complete, another person should check each squaring and total.

    (l) The Preamble details, particularly references to BS Codes, etc., should be checked and any outdated or obsolete references amended. Furthermore, when measured works specifically refer to a particular Preamble clause, it should be checked that they relate to the correct item.

(m) Careful reading over and a final editing of the typed draft bill by an experienced quantity surveyor will provide the last opportunity for corrections prior to printing and may reveal errors which have been missed in the earlier checks.

Differences of technique between individual practices or surveyors should not automatically be assumed to be errors. Moreover, contrasting circumstances and differing customs mean that not every eventuality can be anticipated. Therefore, each of the examples in this book, although based on what is considered to be good practice, should be regarded as offering one alternative and not necessarily as the definitive answer to every possible circumstance.

2

Principles of measurement

The worked examples in this book are based on the traditional dimension paper format and Figure 2.1 shows a typical example of double column dimension paper. For the purposes of this book only the lefthand side of the sheet has been used, to facilitate the use of explanatory notes and a commentary alongside.

Today the use of traditional dimension paper is almost restricted to study and examination purposes, nevertheless it is a convenient vehicle for a textbook on the subject of measurement and the principles, methods and techniques described can be easily adapted to alternative approaches of taking-off.

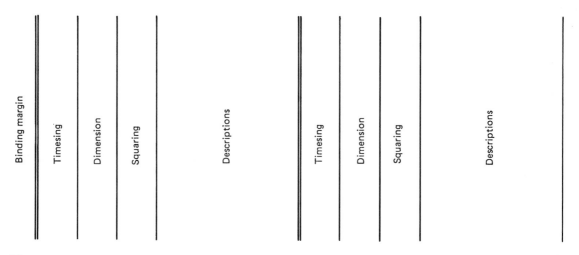

Figure 2.1   *Traditional double column dimension paper*

Each dimension sheet should be titled with at least the name and reference of the project together with the main section of work and numbered. The first sheet should also have the date, the name of the taker-off and the references of the drawings used.

Timesing column

If similar items have the same measurements then, to avoid rewriting, the appropriate measurements may be mutiplied or 'timesed' by writing the timesing figure in the timesing column and separating it from the measurements by a diagonal stroke. Any item 'timesed' may be 'timesed' again, with each multiplier multiplying everything to the right of the diagonal stroke, see Figures 2.2 and 2.3.

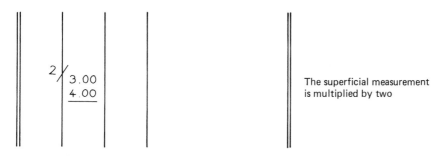

Figure 2.2

The superficial measurement
is multiplied by two

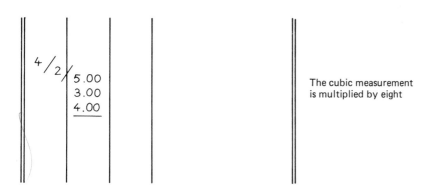

Figure 2.3

The cubic measurement
is multiplied by eight

In some circumstances timesing of previously timesed measurements is not possible but where additional items occur the dimensions may be increased by 'dotting on' or adding as shown in Figure 2.4.

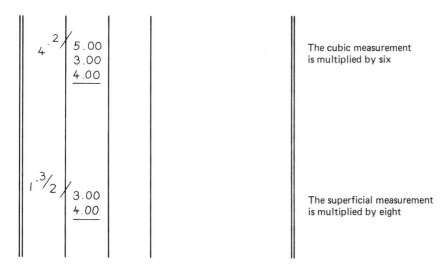

Figure 2.4

The cubic measurement
is multiplied by six

The superficial measurement
is multiplied by eight

Dimension column

General Rule 4.1. requires that dimensions in descriptions follow the sequence: length, width (or breadth), height (or depth). Although the order of dimensions will not affect the calculation of a measurement it is good practice, in an effort to be consistent, to book in the same order. The actual dimensions of an item are entered in the dimension column and it should be noted that all dimensions are expressed in one of the forms shown in Figures 2.5–2.10.

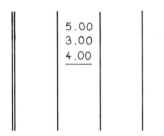

These dimensions indicate a cubic measurement 5.00 metres long, 3.00 metres wide and 4.00 metres deep. The product of the dimensions will be inserted later, adjacent to the figures in the squaring column

Figure 2.5   *Cubic measurements*

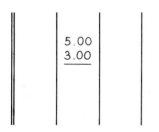

These dimensions indicate a square measurement 5.00 metres long and 3.00 metres wide. The product of the dimensions will be inserted later adjacent to the figures in the squaring column

Figure 2.6   *Square or superficial measurements*

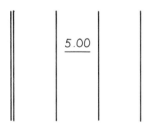

This dimension indicates a linear measurement of 5.00 metres

Figure 2.7   *Linear measurements*

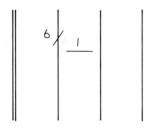

This dimension indicates six in number

Figure 2.8   *Enumerated items*

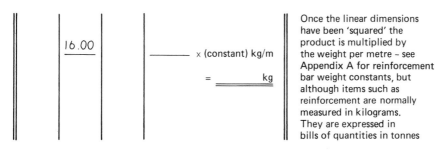

Once the linear dimensions have been 'squared' the product is multiplied by the weight per metre – see Appendix A for reinforcement bar weight constants, but although items such as reinforcement are normally measured in kilograms. They are expressed in bills of quantities in tonnes

Figure 2.9 *Weight – if not given, it may be calculated as shown in the figure*

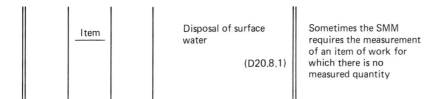

Disposal of surface water

(D20.8.1)

Sometimes the SMM requires the measurement of an item of work for which there is no measured quantity

Figure 2.10 *Items*

Irregular figures

Occasionally it is necessary to measure the areas of triangles, circles and the like and these can be recorded in a number of ways, as illustrated below. However, as previously mentioned, care must be taken when writing fractions to avoid any possible confusion with timesing, as shown in Figure 2.11.

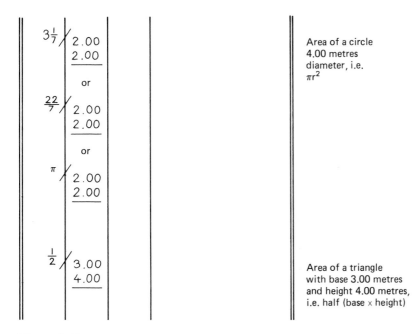

Area of a circle 4.00 metres diameter, i.e. πr^2

Area of a triangle with base 3.00 metres and height 4.00 metres, i.e. half (base × height)

Figure 2.11

Squaring column

The product of the timesing and dimension columns are recorded here prior to being billed or abstracted. The taker-off should not attempt to square dimensions at the same time as taking-off as this could lead to mistakes being made and is a potential waste of resources; this work is usually carried out by a junior member of staff.

Description column

The written description of each item is recorded in the description column, and the information given must include the information necessary to enable the item to be priced accurately.

When the description is dimensioned it is recommended that, whenever practicable, the dimensions be stated at the beginning and not part way through, e.g. 75×225 mm sawn softwood floor members. This allows easier editing, billing and pricing, and it is recommended that the description begins opposite the first dimension as shown in Figure 2.12.

The righthand side of the description column is used for sundry preliminary calculations, explanatory notes and references with regard to the location of the measured work and is termed the 'waste'. Again it is recommended that any waste calculations are carried out prior to and kept clear of descriptions. Waste calculations are written to three decimal places and the results reduced to the nearest 10 mm (General Rule 3.2) before being transferred to the dimension column.

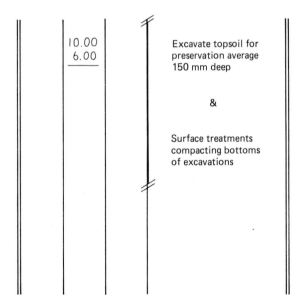

Figure 2.12

The ampersand

When dimensions are to be repeated for other items this is indicated by use of the ampersand sign as shown in Figure 2.12, thereby preventing repetition. However, care must be taken when combining linear with superficial items or superficial with cubic items.

Deductions

Measurements should always be carried out 'working from the whole to the part', i.e. measure overall and then make any deductions if necessary.

Deductions of previously measured quantities may be made either in the squaring column or, more conventionally, in the description column as shown in the following examples. In either case, the item must be preceded by the word 'Deduct' (or its abbreviated form 'Ddt'); and to ensure that only the intended items are properly deducted the next item should be preceded by the word 'Add' for the avoidance of any doubt. Furthermore, it is also good practice to highlight the respective instructions for clarity. Figure 2.13 illustrates two alternatives.

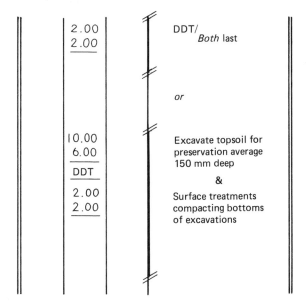

Figure 2.13

From experience it has been found that if only one or two deductions are necessary then an adjustment in the dimension column has much to commend it although in the following examples the more traditional approach has been adopted. However, if there are a number of deductions then it is recommended that the description deduction alternative is more suitable.

Brackets

Traditionally a bracket is used whenever more than one measurement relates to a description or group of descriptions, see Figure 2.13. However, to avoid any possible confusion and certainly to improve presentation, the use of brackets irrespective of the number of dimensions or descriptions is recommended.

Descriptions

Descriptions in the bills of quantities must comply with the requirements of SMM7 (see General Rules Clause 4), although it should be remembered that more information should be given where necessary in order to define the precise nature and extent of the work.

Descriptions need to be adequate but, at the same time, need to be clear and concise and whilst the use of traditional prose is not precluded the examples contained in this book have been built up following the order set out in the tabulated rules of the SMM; that is to say, that each description shall include one descriptive feature from each of the first three columns in the classification table and as many of the descriptive features in the fourth column (ignoring numbering the unit of measurement column) as are applicable, together with the necessary information from the supplementary rules. The aim of this standardization is to encourage consistency of description which inevitably leads to improved communication.

Extra over

Certain items are measured as extra over the item of work in which they occur, in which case the estimator will price the additional cost involved as to some extent this item will have been previously measured. An example of this would be bends on rainwater goods.

Written short

Items which are to appear inset in the bills of quantities, usually described as 'written short', are entered in the same manner as other descriptions. The letters WS/ are used as a prefix to the description of the item in question.

Abbreviations

In order to save space and time many descriptions are abbreviated. There are no fully accepted standard abbreviations although a list of those more commonly used is set out in Appendix 1. This is by no means exhaustive as, in practice, the list will be greatly extended by the shortening of familiar words in the descriptions, provided the meaning is clear and unambiguous, and in addition individual offices may have their own preferences.

Schedules

In addition to providing design information, a schedule can be produced to facilitate the taking-off process. Schedules may be used to collect all relevant specification information in a concise tabulated form for ease of reference to assist measurement or they may be used for recording measurements and, in effect, may become the taking-off. The use of schedules for items which repeat themselves throughout the building is advocated as a means of saving time and, due to the more systematic approach, reducing the possibility of errors. Schedules are particularly appropriate in the measurement of doors, windows, finishings and drainage, although if only small numbers are involved any advantages would be lost.

Drawn information

General Rules Clause 5 refers to four types of drawn information:

1. location drawings
2. component drawings
3. dimension diagrams
4. schedules

With regard to dimension diagrams, sometimes rather than a lengthy and possibly ambiguous dimension description a sketch included in a bill of quantities may be more appropriate. It has the advantage of showing clearly the shape and dimensions of the work, although a bill diagram should not replace an item otherwise required to be measured.

Deemed to be included items

Items which are deemed to be included are mentioned in the Coverage rules of SMM7 and indicate that an allowance is to be made by the estimator for these specific items which are not required to be separately measured.

Query sheets

During the taking-off process problems may arise involving the interpretation of drawings, clarification of the specification or additional information which will require a decision from the architect or engineer. In the event of this a query sheet should be produced by the taker-off. It is strongly advised for examination purposes that students incorporate a query sheet along similar lines to that included in Appendix 2, perhaps even answering the query themselves.

Taking-off procedure

1. Head the first sheet with the name of the project, date, name of taker-off, drawing numbers and section of work measured and adopt a similar system of identification for subsequent sheets.
2. Do not crowd dimensions and write legibly. Neat presentation is paramount.
3. Signpost dimensions, give locational information and sub-headings in the waste column, but keep clear of descriptions.
4. Approach drawings in a standard sequence, e.g. starting at the top lefthand corner and follow a clockwise pattern.
5. Ensure that all items are measured by taking-off in a logical sequence and that dimensions follow on.
6. Cross reference where dimensions are added back, when to take notes have been cleared or where schedules are used.
7. If necessary, correct dimensions in the right place; this is normally where they are first recorded, but if this is not possible use cross references.
8. Line through specification notes, taking off list items, query lists, dimensions and schedules when items have been measured.
9. Use 'to take' notes when information is not to hand or as an aide memoire and check that all 'to take' notes are cleared before completion.
10. Check through drawings, query lists, specifications and schedules on completion to ensure that all items have been taken.

Girthing

Perhaps the most fundamental of tasks in taking-off is to establish the girth and centre line ($\mathcal{C}$) of a building or wall. These may be calculated as shown in Figures 2.14–2.17.

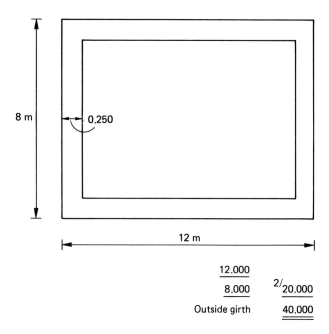

$$\frac{12.000}{8.000} \quad \frac{^{2/}20.000}{}$$

Outside girth   40.000

Figure 2.14   *Plan*

To obtain the ℄ of the above, for each corner deduct
half the thickness of the wall in each direction

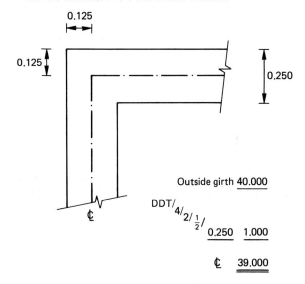

Outside girth 40.000

DDT/
    ⁴/₂/ ½ /
        0.250   1.000

℄   39.000

Figure 2.15   *Plan of corner*

In the event of a re-entrant corner the girth and the centre line remain the same as if the
building had been a complete rectangle, see Figure 2.16.

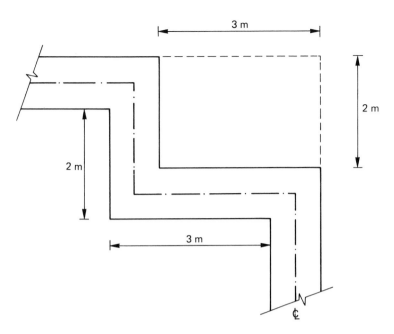

Figure 2.16   *Plan of re-entrant corner*

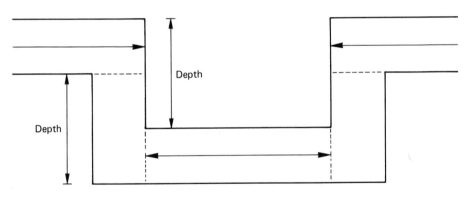

Figure 2.17   *Plan of re-entrant*

The situation is different if the re-entrant occurs somewhere along the wall other than at a corner, see Figure 2.17. In this case add twice the depth of the re-entrant to the girth or centre line.

In conclusion, measurement is basically a practical skill which involves the following key factors:

1. A disciplined, logical and consistent approach.
2. A sound knowledge of construction technology to facilitate the interpretation of drawings and three-dimensional thinking.
3. The ability to apply the rules of measurement as laid down in the SMM or other recognized form of measurement.
4. Sufficient understanding and knowledge of mensuration techniques to calculate measurements.
5. The application of checks at all stages of the work.

Foundations

Introduction

Foundations or sub-structure is the term used to indicate all supporting building work in the ground below a specific demarcation line, usually the damp proof course and/or damp proof membrane.

In this first example all descriptions are written out in full, but it is usual for these to be abbreviated in practice.

Information required

SMM D20.P1 lists information which should accompany the bills of quantities. The following information is required before measurement can commence:

1. full specification
2. site survey report including the results of trial pits and/or bore hole investigations
3. plan of foundations complete with sections and levels

With regard to levels, the following information is required:

1. existing ground levels
2. floor level of proposed building
3. level showing depth of foundations
4. finished ground levels

Existing ground levels will be found on the 'site survey' drawings which should also show the position of trial pits and/or bore holes including date when inspected. In addition, the positions and types of trees, existing buildings, services, paved areas and other features which may affect the measurement should be included.

Finished floor levels and levels showing depths of foundation will be found or can be calculated from the architect's or engineer's drawings. Finished ground levels will be found on the external works drawings.

Measurement

Excavation

Excavation oversite to remove topsoil is one of the first operations carried out. Topsoil may be kept on site in temporary spoil heaps for use later in the execution of the external works. The depth of topsoil is obtained from the site survey investigations and if it is to be preserved it is measured in square metres, see SMM D20.2.1.

The oversite excavation is measured to the extreme dimensions of the building's foundation, even though in practice the whole site may be stripped of topsoil in one operation. If there is no topsoil or if the topsoil is not to be retained then the resultant excavation to reduced level is measured in cubic metres and described as such.

The oversite excavation must be completed before the measurement of excavation of trenches or pits of which the depth is calculated from this reduced level (also known as the commencing level of excavation), and if this is more than 250 mm below existing ground level, this must be stated.

The description and measurement rules in SMM7 are based on mechanical forms of excavation. Therefore the classifications and maximum depths must be given to provide sufficient information to enable the estimator to price the excavation using the most economic type of plant for the particular project. The maximum depth given can be for each unit or group of units (see Code of Procedure).

Excavation is measured to the shape shown on the drawings and the estimator must allow in his pricing for bulking of the soil and additional space to accommodate earthwork support, see SMM D20.M3.

Earthwork support

Earthwork support is to be measured to the full depth of all faces of excavation whether or not required to all vertical faces exceeding 0.25 metres high, see SMM D20.7.*.*.*.M9. It is up to the contractor to decide whether or not earthwork support is required to any particular excavation.

Earthwork support is classified by the maximum depth of the excavation and the distance between the opposing face, i.e. the distance the supporting struts have to span. However, it should be noted that the word 'opposing' does not necessarily mean opposite. For example, in the plan shown in Figure 3.1 all the earthwork support is classified as 'Distance between opposing faces ⩽ 2.00 m' even though some opposing faces exceed 2.00 m. There would be no earthwork support measured in the categories 2.00–4.00 m (faces Y–Y) or > 4.00 m (faces X–X) between opposing faces.

Where earthwork support to oversite excavation or reduced level excavation coincides with the external face of a trench and does not exceed 0.25 m high it is in practice measured and given the same 'distance between opposing faces' classification as the trenches. Where it exceeds 0.25 m high it should be measured and given the appropriate classification, e.g. distance between opposing faces > 4.00 m.

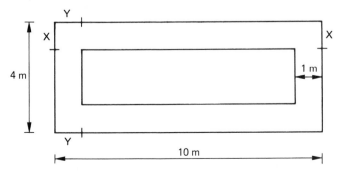

Figure 3.1 *Plan of foundation trenches*

Concrete in foundations

Concrete in foundations is measured to the exact dimensions shown on the drawings but must be distinguished as between being 'poured on or against earth or unblinded hardcore' so that the contractor can make due allowance for any concrete loss due to irregularities.

Brickwork

Generally brickwork is measured in square metres and the thickness is stated, as shown in Figures 3.2–3.4. This is usually given in numbers of bricks and is measured on the bed of the brick at right angles to the face of the wall.

The actual size of standard bricks is $215 \times 102.5 \times 65$ mm, but the nominal size is $225 \times 112.5 \times 75$ mm, which includes a 10 mm allowance for a mortar joint. This information will be used to calculate wall, pier and chimney stack widths and thicknesses in the following worked examples.

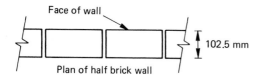

Figure 3.2   *Plan of half brick wall*

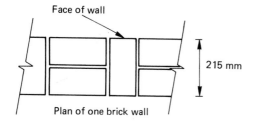

Figure 3.3   *Plan of one brick wall*

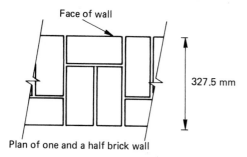

Figure 3.4   *Plan of one and a half brick wall*

Faced brickwork

Exposed faces of brick walls will usually be built using facing bricks and the joints pointed. To allow for the soil backfilling of trenches to settle, one or preferably two courses of facing brickwork are taken below the finished ground level. In the following worked examples two courses are taken below the finished ground level.

Flow chart for foundation measurement

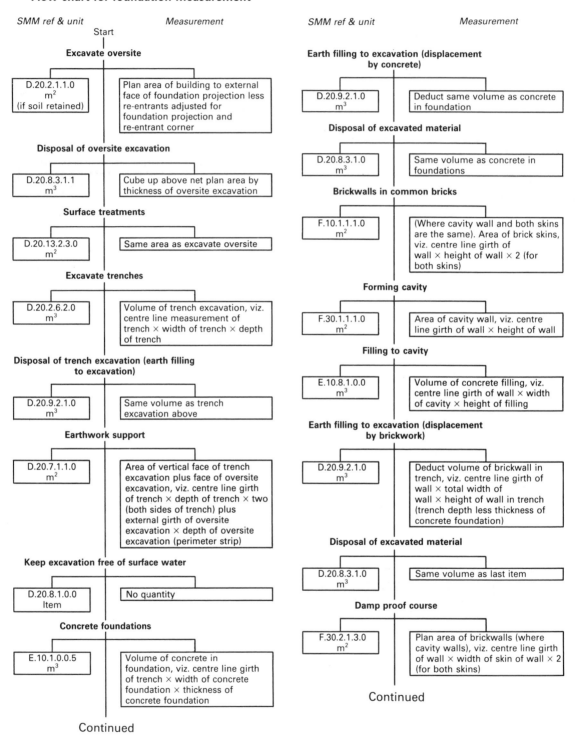

SMM ref & unit	Measurement
Start	

Excavate oversite

| D.20.2.1.1.0 m² (if soil retained) | Plan area of building to external face of foundation projection less re-entrants adjusted for foundation projection and re-entrant corner |

Disposal of oversite excavation

| D.20.8.3.1.1 m³ | Cube up above net plan area by thickness of oversite excavation |

Surface treatments

| D.20.13.2.3.0 m² | Same area as excavate oversite |

Excavate trenches

| D.20.2.6.2.0 m³ | Volume of trench excavation, viz. centre line measurement of trench × width of trench × depth of trench |

Disposal of trench excavation (earth filling to excavation)

| D.20.9.2.1.0 m³ | Same volume as trench excavation above |

Earthwork support

| D.20.7.1.1.0 m² | Area of vertical face of trench excavation plus face of oversite excavation, viz. centre line girth of trench × depth of trench × two (both sides of trench) plus external girth of oversite excavation × depth of oversite excavation (perimeter strip) |

Keep excavation free of surface water

| D.20.8.1.0.0 Item | No quantity |

Concrete foundations

| E.10.1.0.0.5 m³ | Volume of concrete in foundation, viz. centre line girth of trench × width of concrete foundation × thickness of concrete foundation |

Continued

SMM ref & unit	Measurement

Earth filling to excavation (displacement by concrete)

| D.20.9.2.1.0 m³ | Deduct same volume as concrete in foundation |

Disposal of excavated material

| D.20.8.3.1.0 m³ | Same volume as concrete in foundations |

Brickwalls in common bricks

| F.10.1.1.1.0 m² | (Where cavity wall and both skins are the same). Area of brick skins, viz. centre line girth of wall × height of wall × 2 (for both skins) |

Forming cavity

| F.30.1.1.1.0 m² | Area of cavity wall, viz. centre line girth of wall × height of wall |

Filling to cavity

| E.10.8.1.0.0 m³ | Volume of concrete filling, viz. centre line girth of wall × width of cavity × height of filling |

Earth filling to excavation (displacement by brickwork)

| D.20.9.2.1.0 m³ | Deduct volume of brickwall in trench, viz. centre line girth of wall × total width of wall × height of wall in trench (trench depth less thickness of concrete foundation) |

Disposal of excavated material

| D.20.8.3.1.0 m³ | Same volume as last item |

Damp proof course

| F.30.2.1.3.0 m² | Plan area of brickwalls (where cavity walls), viz. centre line girth of wall × width of skin of wall × 2 (for both skins) |

Continued

Flow chart for foundation measurement – continued

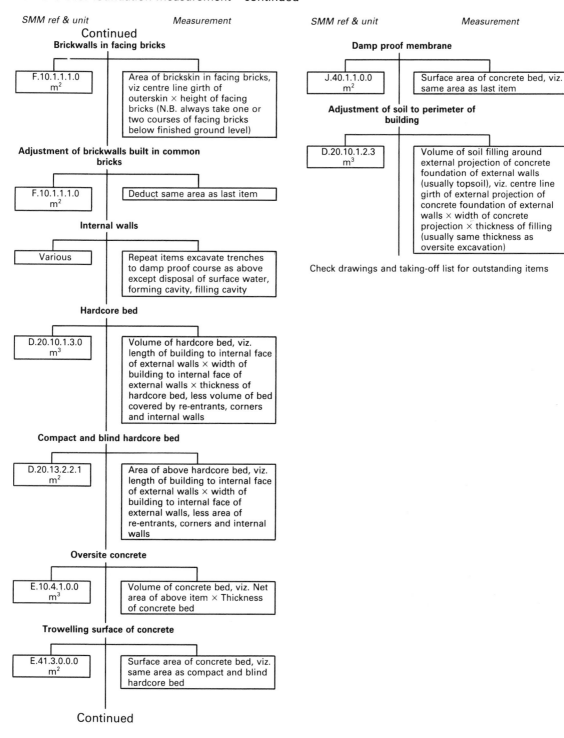

SMM ref & unit **Measurement**

Continued

Brickwalls in facing bricks

F.10.1.1.1.0
m²
— Area of brickskin in facing bricks, viz centre line girth of outerskin × height of facing bricks (N.B. always take one or two courses of facing bricks below finished ground level)

Adjustment of brickwalls built in common bricks

F.10.1.1.1.0
m²
— Deduct same area as last item

Internal walls

Various
— Repeat items excavate trenches to damp proof course as above except disposal of surface water, forming cavity, filling cavity

Hardcore bed

D.20.10.1.3.0
m³
— Volume of hardcore bed, viz. length of building to internal face of external walls × width of building to internal face of external walls × thickness of hardcore bed, less volume of bed covered by re-entrants, corners and internal walls

Compact and blind hardcore bed

D.20.13.2.2.1
m²
— Area of above hardcore bed, viz. length of building to internal face of external walls × width of building to internal face of external walls, less area of re-entrants, corners and internal walls

Oversite concrete

E.10.4.1.0.0
m³
— Volume of concrete bed, viz. Net area of above item × Thickness of concrete bed

Trowelling surface of concrete

E.41.3.0.0.0
m²
— Surface area of concrete bed, viz. same area as compact and blind hardcore bed

Continued

SMM ref & unit **Measurement**

Damp proof membrane

J.40.1.1.0.0
m²
— Surface area of concrete bed, viz. same area as last item

Adjustment of soil to perimeter of building

D.20.10.1.2.3
m³
— Volume of soil filling around external projection of concrete foundation of external walls (usually topsoil), viz. centre line girth of external projection of concrete foundation of external walls × width of concrete projection × thickness of filling (usually same thickness as oversite excavation)

Check drawings and taking-off list for outstanding items

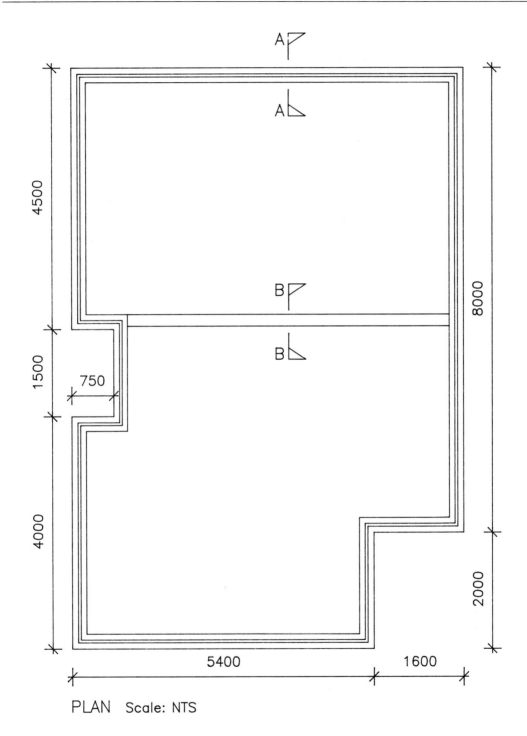

PLAN   Scale: NTS

FOUNDATIONS

SPECIFICATION
1. Topsoil: 150 mm deep — to be excavated & deposited in spoil heaps 50 m from excavation.
2. All surplus soil taken to a tip provided by the contractor.
3. No ground water.
4. Concrete: a) Foundations — 15N/20 mm
   b) Oversite & cavity fill — 20N/20 mm
5. Brickwork: a) Generally to be Class B engineering bricks in cement mortar — (1:3)
   b) Facing bricks to be Redland multicoloured in cement mortar — (1:3) pointed with a flush joint as work proceeds.
   c) Built in stretcher or English bond.
6. Wall tiles: Mild steel vertical twisted type — to BS.1243 Type 3 — 2 per sq. metre.
7. DPC: Hyload pitch polymer bedded in cement mortar (1:3) lapped 150 mm.
8. DPM: 1000 gauge polythene lapped 150 mm at all joints.
9. Hardcore: Broken brick or stone — blinded with sand.

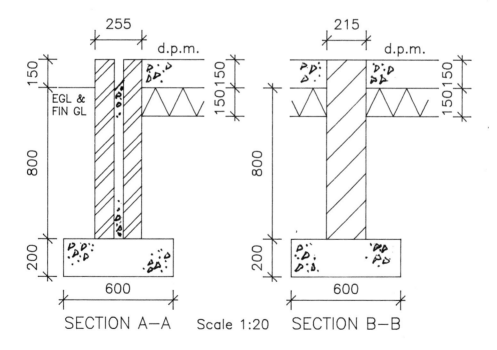

SECTION A—A   Scale 1:20   SECTION B—B

FOUNDATIONS

Foundations 1.

Taking Off List.

1. Excavate vegetable soil and disposal of soil.
2. Surface treatments
3. External walls
   a. Excavate trenches and disposal of soil.
   b. Earthwork support.
   c. Keep excavations free of surface water
   d. Concrete in foundations and adjust disposal of soil.
   e. Brickwork up to damp proof course and ditto.
      i. Outer and inner skins
      ii. Formation of cavity.
      iii. Concrete filling to cavity
   f. Damp proof course
   g. Facing brickwork to external skin and adjustment of common brickwork.
4. Internal walls repeat a-f above.
5. Hardcore bed.
6. Blind hardcore bed
7. Oversite concrete
8. Trowelling surface of concrete
9. Damp proof membrane.
10. Adjustment of top soil to perimeter.

Page Nrs.

1	11
2	12
3	13
4	14
5	15
6	16
7	17
8	18
9	19
10	20

(left margin, vertical) Drawing Number.   Name of Project.   date   Name 1   Name.

Foundations 2.

Excavate oversite.

5·400	×	8·000	
1·600		2·000	
External dimensions 7·000		10·000	
+Concrete projection			
trench 0·600			
– brick wall 0·255			
2/½/ 0·345 : 0·345		0·345	
7·345		10·345	

Name 2   date   Name of Project

Side notes indicate section measured. (See taking-off list.)

Waste calculations to three decimal places. (See SMM General Rules 3.2.)
The convention of noting horizontal dimensions followed by vertical dimensions is used here.

The oversite exacavation must be taken to the extreme dimensions necessary to enable the foundations to be dug. Therefore the projection of the concrete foundation is added to the external brickwork dimensions.

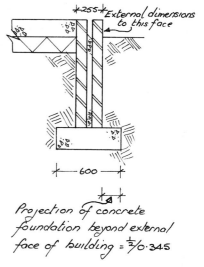

Projection of concrete foundation beyond external face of building = ²/0·345

It is necessary to add this same dimension for the concrete projection on the other face of the building.

Foundations 3

Excavate oversite (Ctd.)

7·35	Excavating topsoil for preservation average 150mm deep	SMM D20.2.1.1.0

&

The ampersand is used to save writing the dimensions again.

Disposal excavated material on site in spoil heaps average 50m from excavations Cube × 0·15 = _____ m³	SMM D20.8.3.1.1

The instruction given after a description converts the superficial measurement into a cubic measurement.

&

Surface treatments compacting bottoms of excavations.

SMM D20.13.2.3.0

Re - entrant.

Side notes to identify particular points or areas.

```
                    1·500
-concrete projection
         20/½/0·345 = 0·345
                    1·155
```

Waste calculation set out to enable complete checking of dimensions.

1·60	Ddt excavating topsoil (corner as before	SMM D20.2.1.1.0
2·00		
0·75	(re-entrant	The 'Golden Rule' of measurement is 'measure overall and then adjust'. It is necessary to adjust the topsoil excavation, etc., for the wants.
1·16		

Name of Project

date

Name 3

Foundations 4.

Excavate oversite (Ctd)

1·60 2·00 0·75 1·16	*Ddt* disposal excavated (corner material on site in spoil heaps as before Cube × 0·15 = _____ m³ (re-entrant

SMM D20.8.3.1.1

The instruction Deduct (*Ddt*) is written in front of the description of the item which has to be reduced.

&

Ddt surface treatments compacting as before

SMM D20.13.2.3.0

The term as before (usually written a.b.) is used to save writing full descriptions for items fully described previously.

₵ = Centre Line calculation. External dimensions from Foundations 2.

External Walls
Trench ₵

```
              7·000
             10·000
          2/ 17·000
           =  34·000
```

Re-entrant ²/0·750 = 1·500

External girth of brickwork = 35·500

− passings ⁴/²/²/0·255   1·020

₵ = 34·480

Trench depth

Existing ground level to top of concrete foundation } 0·800

+ Concrete foundation   0·200

1·000

− Oversite excavation   0·150

Trench depth   0·850

The total depth of foundation trench is calculated from the drawing, i.e. existing ground level to underside of foundation concrete less the thickness of the topsoil which has already been measured.

Name 5   date   Name of Project.

Foundations 5

External Wall (Ctd)

34·48
0·60
0·85
5·40
0·60
0·85

Excavating trenches width > 0·30 m maximum depth ≤ 1·00m

⎰internal
⎱wall

&

SMM D20.2.6.2.0

See Foundations 13 for waste calculations for internal wall.

Filling to excavation average thickness >0·25m arising from excavations.

SMM D20.9.2.1.0

Part of the excavated soil is required to be returned to the trench and part disposed of elsewhere. The conventional way of measuring this item is by filling the excavated material back into the trench and then adjusting for the volume of concrete and brickwork in the trench later.

Sometimes it may be more convenient to dispose of the soil elsewhere and adjust for any soil to be returned to the trench later.

Earthwork support.
External girth of trench
₵ girth   34·480
+passings 4/2/½/0·600   2·400
girth of oversite strip 36·880

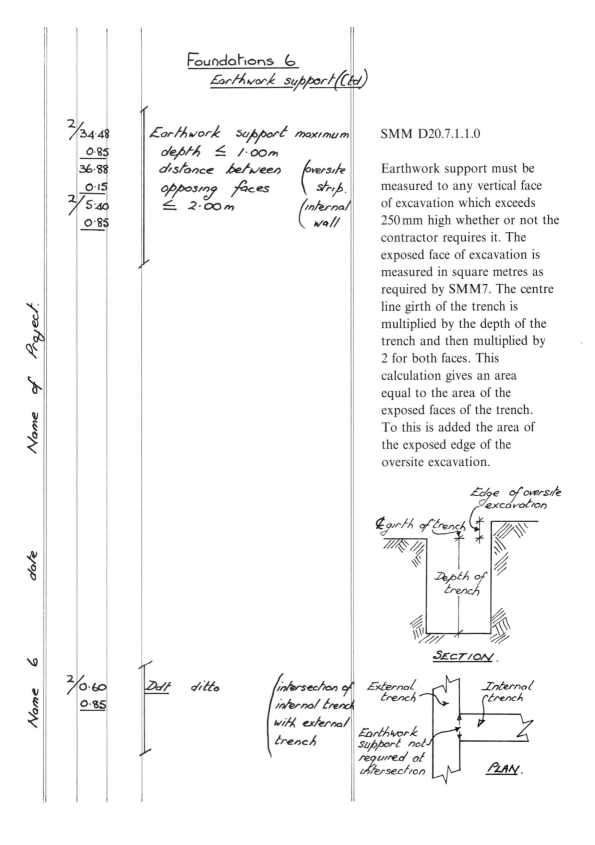

Foundations 6
Earthwork support (Ctd)

2/34·48	Earthwork support maximum
0·85	depth ≤ 1·00m
36·88	distance between {oversite
0·15	opposing faces { strip.
2/5·40	≤ 2·00m {internal
0·85	{ wall

SMM D20.7.1.1.0

Earthwork support must be measured to any vertical face of excavation which exceeds 250 mm high whether or not the contractor requires it. The exposed face of excavation is measured in square metres as required by SMM7. The centre line girth of the trench is multiplied by the depth of the trench and then multiplied by 2 for both faces. This calculation gives an area equal to the area of the exposed faces of the trench. To this is added the area of the exposed edge of the oversite excavation.

Edge of oversite excavation

℄ girth of trench

Depth of trench

SECTION.

2/0·60	Ddt ditto
0·85	

{intersection of
{internal trench
{with external
{trench

External trench

Internal trench

Earthwork support not required at intersection

PLAN.

Foundations 7
Surface Water

Item	Disposal, surface water	SMM D20.8.1.0.0

Surface water is water which runs over the surface of the ground and collects in the excavations (see SMM D20.8.1 −D9), the disposal of which is given as an item whenever any excavation is measured. The estimator assesses the cost based on the amount of excavation, e.g. open trenches, and when the foundations are likely to be built, e.g. winter or summer.

Concrete foundations

34·48	In situ concrete (15N/20mm) foundations poured on or against earth or unblinded hardcore	SMM E10.1.0.0.5
0·60		
0·20		
5·40	(internal	
0·60	wall)	
0·20		

The phrase 'poured, on or against earth etc' is necessary because work is measured net as fixed in position (see SMM General Rules 3.1) and in practice the trench may not be excavated as a perfect rectangle; therefore the estimator has to make an allowance in the price to cover such eventualities.

Name of Project.

Name 7   date

<u>Foundations 8</u>
<u>Concrete foundations(Ctd)</u>

34.48		
0.60		
0.20		
5.40		
0.60		
0.20		

<u>Ddt</u> filling to excavations as before

            &   (internal wall

SMM D20.9.2.1.0

Instructions Add and Ddt are emphasized by underlining.

<u>Add</u> disposal excavated material off site

SMM D20.8.3.1.0

<u>Brickwork Height.</u>

Brickwork in trench   0.800
Brickwork above ground)
level to damp proof }
course   ) 0.150
               0.950

The required height of brickwork is from the top of the foundation concrete to the damp proof course.

2/34.48		
0.95		

Walls half brick thick, vertical in Engineering bricks in cement mortar (1:3) in stretcher bond.

SMM F10.1.1.1.0

The component parts of cavity walls are each measured separately, e.g. brickwalls, cavity and cavity filling.

The word 'vertical' can be omitted 'as work is deemed to be vertical unless otherwise described'. (See SMM F.10-D3.)

Name of Project.

date

Name 8

Foundations 9
        Brickwork (Ctd)

Where the inner and outer skins of hollow walls are built of the same material, are the same width and are central on the concrete foundation, the walls can be measured on the centre line girth. If the skins are built of different materials or are of different thicknesses then the centre line of each skin must be calculated.

The walls are measured overall ignoring the fact that part of the outer skin is built in facing bricks. The adjustment will be made later.

The area of the brick skins of the hollow walls is calculated by multiplying the centre line girth by the height by 2 for both skins.

Cavity
wall thickness   0.255
−brick walls ²/.1025   0.205
cavity width   0.050

34.48	Forming cavities in hollow walls 50 mm wide, stainless steel wall ties as specified 2 per square metre.
0.95	

SMM F30.1.1.1.0

SMM F30-S2 requires the spacing of wall ties to be given. In this case the description could be written as 'wall ties … 1.00 m apart horizontally and 0.50 m vertically, staggered'.

Name of Project.

date

Name 9

Foundations 10
Brickwork (Ctd)

Cavity filling.
height of cavity 0.950
- distance below damp
proof course ⎱ 0.150
cavity filling 0.800

34.48	In situ concrete (20N/20mm) filling to hollow walls thickness ≤ 150mm
0.05	
0.80	

SMM E10.8.1.0.0

The thickness given in the description enables the estimator to calculate the site labour required to place 1 m³ of concrete.

Adjustment of
filling to trench
Trench depth 0.850
- concrete foundation 0.200
~~0.750~~
0.650

If an error is made in a side cast calculation, this should be crossed through and the correct figures written underneath.

34.48	Ddt filling to excavations as before.
0.26	
0.75 NIL	
34.48	
0.26	
0.65	
5.74	&
0.22	(internal wall
0.65	

SMM D20.9.2.1.0

If an error is made when writing dimensions, the incorrect dimension should be nilled as shown and the correct dimension written underneath.

Add disposal excavated material off site.

SMM D20.8.3.1.0

The dimension 0.26 is the width of the cavity wall 0.255.

Name of Project.

date

Name 10

Foundations 11

Damp proof course

2/	34·48	
	0·10	
	5·74	
	0·22	

Damp proof course width ≤ 225 mm, horizontal single layer of 'Hyload' pitch polymer lapped 150 mm at all passings, bedded in cement mortar (1:3)   (*internal wall*)   (Meas nett)

SMM F30.2.1.3.0

Facing brick adjustment

Outerskin built in facing bricks

Finished Ground Level

Common brickwork

SECTION

All brickwork has been measured in common bricks. The outerskin shown hatched on above section can be seen and this part of the wall is built in facing bricks and the joints pointed. The facing bricks are usually extended one or two courses below ground level to take account of irregular ground lines and/or settlement of filling.

Name of Project.

date

Name 11

Foundations 12
Facing brick adjustment (Ctd)

$\cancel{L}$ of outerskin
External girth of walls 35·500
—passings 4/2/½/0·1025 0·410
$\cancel{L}$ of outerskin 35·090

It is necessary to calculate the centre line girth of the outerskin in order to find the area of faced brickwork.

Height
Ground level to damp
proof course ⎰ 0·150
Below ground level 0·150
0·300

35·09	Walls facework one side half brick thick, vertical in Redland Multi Coloured Facing Bricks in cement mortar (1:3) in stretcher bond and pointing with a flush joint as work proceeds.
0·30	

SMM F10.1.1.1.0

&

Ddt walls half brick thick in Engineering bricks as before.

SMM F10.1.1.1.0

The description for a repeat item or deduction need only be full enough to enable the worker-up to recognize the item to which it relates.

Name of Project.

date

Name 12

Name

Foundations 13
        Internal Wall

Depending on the form of construction, the internal wall foundation may be measured with the external wall foundation, but here it is measured separately to enable the student to concentrate on one section of the foundation at one time.

One system is to add these dimensions back in a coloured ink as this saves making notes against each set of dimensions (as done in this example) and enables the worker-up to check that all items are measured. The term *added back* means that the taker-off will add on to previously booked dimensions, thus saving working-up time, taking-off time and paper in repeating descriptions.

Length of brickwall
overall dimension   7·000
– re-entrant   0·750
                      6·250
– external brickwalls
         2/0·255: 0·510
length of 1B wall   5·740

Length of trench etc
length of 1B wall   5·740
– concrete projection⎱
of external wall   ⎰
foundation   ⎰
         2/½/0·345: 0·345
length of trench   5·395

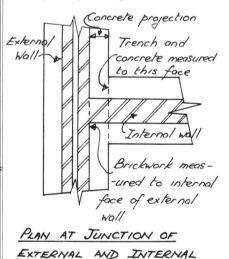

External Wall →

Concrete projection

Trench and concrete measured to this face

Internal wall

Brickwork meas-
-ured to internal
face of external
wall

PLAN AT JUNCTION OF
EXTERNAL AND INTERNAL
WALLS

Foundations 14
Internal Wall (Ctd)

When measuring internal wall foundations it should be noted that the length/girth of the brickwork is longer than the length/girth of the trench and concrete. This is due to the projection of the concrete foundation beyond the brick face at the junction of walls.

Some depth as external wall.

The dimensions for the internal wall are added back to Foundations 5 et seg. except :

5·74	Walls one brick thick,
0·95	vertical in Engineering
	bricks in cement mortar
	(1:3) in english bond

SMM F10.1.1.1.0

Hardcore bed

External dimensions 7·000 × 10·000
−external walls ²∕₆ 255:0·510   0·510
6·490   9·490

6·49	Filling to make up levels
9·49	average thickness ≤ 0·25m
0·15	hardcore obtained off
	site

SMM D20.10.1.3.0

The average thickness of the filling must be calculated to place the filling in categories according to SMM D20.10.1 or 2. In this case the filling is the same thickness throughout.

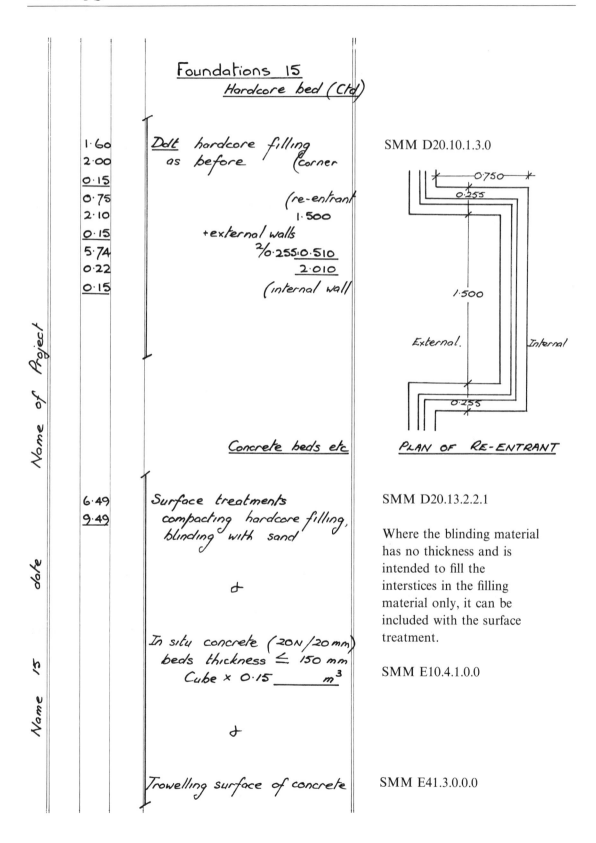

Foundations 15
   Hardcore bed (Ctd)

1·60	<u>Ddt</u> hardcore filling
2·00	as before (corner
0·15	
0·75	(re-entrant
2·10	1·500
0·15	+external walls
5·74	²∕₀·255:0·510
0·22	2·010
0·15	(internal wall

SMM D20.10.1.3.0

0·750
0·255

1·500

External.   Internal.

0·255

PLAN OF RE-ENTRANT

Concrete beds etc

6·49	Surface treatments
9·49	compacting hardcore filling,
	blinding with sand

SMM D20.13.2.2.1

Where the blinding material has no thickness and is intended to fill the interstices in the filling material only, it can be included with the surface treatment.

&

In situ concrete (20N/20mm)
   beds thickness ≤ 150 mm
      Cube × 0·15 _____ m³

SMM E10.4.1.0.0

&

Trowelling surface of concrete

SMM E41.3.0.0.0

Name of Project.

Name 16   date

Foundations 16
Concrete beds etc (Ctd)

1·60	_Ddt_ surface treatments,	SMM D20.13.2.2.1
2·00	compacting hardcore (corner	
0·75	as before	
2·01	(re-entrant	
5·74		
0·22	& (internal wall	

	Ddt in situ concrete	SMM E10.4.1.0.0
	(20 N/20 mm) as before	
	Cube × 0·15 = _____ m^3	

&

	Ddt trowelling surface	SMM E41.3.0.0.0
	of concrete as before	

Damp proof membrane

6·49	Damp proof membrane .	SMM J40.1.1.0.0
9·49	horizontal of 1000 gauge	
	polythene sheet as specified	The extent of laps, the
	laid on concrete.	measurement of which is deemed
		included, should be specified in
		the Preambles section.

1·60	_Ddt_ ditto (corner	SMM J40.1.1.0.0
2·00		
0·75	(re-entrant	
2·01		
5·74	(internal wall	
0·22		

Foundations   17
            Adjustment of top
            soil to perimeter

                        ₵ of projection
External girth of ⎫
    external walls ⎬   35.500
concrete projection
                    2)0.345
                    ──────
                  = 0.173
                    4/2½/0.173 = 0.692
    +passings   ──────
₵ of external concrete ⎫   36.192
        projection   ⎬   ──────

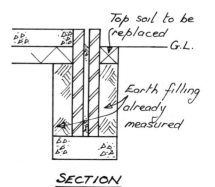

Top soil to be
replaced
        ─── G.L.

Earth filling
already
measured

SECTION

36.19		
0.17		
0.15		

Filling to make up levels
average thickness ≤ 0.25m
obtained from on site
top soil spoil heaps
average 50m from
excavation.

SMM D20.10.1.2.3

Name of Project.   date   Name 17

4

Foundations on a sloping site

Introduction

To achieve an economic design for foundations on a sloping site part of the building will be in an area of cut or reduce level excavation and part in an area of fill, see Figure 4.1.

In addition to the information required as listed in the previous chapter, the foundation plan must show the levels and depths of any step foundations and their position relative to the corners of the building. To enable the calculations to be made it is assumed that the surface of the ground is constant (not necessarily level) between the positions where existing ground levels are taken, even though the surface of the ground undulates.

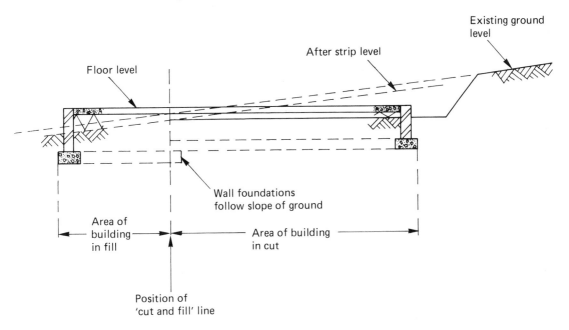

Figure 4.1   *Section – Foundations on a sloping site*

Calculations

During the measurement the following calculations need to be made:
1. Calculate the position of the 'cut and fill' line and plot on the drawing. From this the amount of cut or fill from existing ground levels and the required formation level can be ascertained, see Figure 4.2.

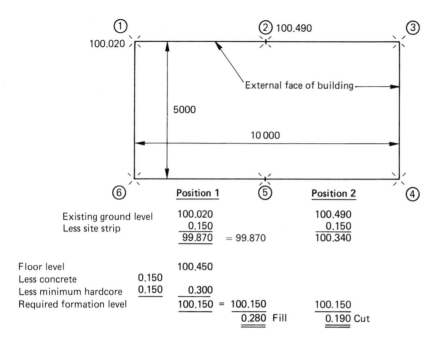

Figure 4.2   *Plan of building*

At station 1, the required formation level is 0.28 m above the after strip level. Therefore 0.28 m is the amount of additional filling required to get to formation level, i.e. the underside of the hardcore bed. Thus, given a hardcore bed of 150 mm, a total depth of filling of 430 mm (0.280+0.150) will be required. At station 2, the required formation level is 0.19 m below the after strip level. Therefore, 0.19 m is the depth of reduce level digging at station 2.

2.   Calculate the position of the cut and fill line and plot on to drawing. It can be seen that one end of the cut and fill line must occur between stations 1 and 2 because station 1 is in fill and station 2 is in cut. Therefore, the position of the cut and fill line can be calculated by interpolation, see Figure 4.3.

3.   Calculate the position of a line (PQ) where the trench excavation exceeds 0.25 , below existing ground level and plot on drawing, see Figure 4.4.

Line occurs where 0.1 m of reduce level excavation occurs (0.250−0.150 site strip=0.100), see Figure 4.5.

4.   Calculate the depth of trenches, including steps, to find the maximum depth of trench excavation, in this case position 5.

After strip level		99.960
− Bottom of standard trench=99.300		
− Step	= 0.225	99.075
Maximum depth of trench		0.885

Therefore description will include '... maximum depth ⩽ 1.00 metres'.

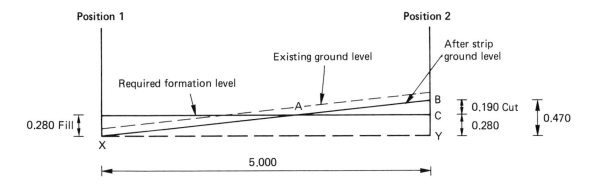

$\triangle$ ABC is similar to $\triangle$ XBY, therefore the distance AC
can be found by interpolation

$$AC = \frac{0.190}{0.470} \times 5.000 = \underline{2.021 \text{ m}}$$

or: 5.000 − 2.021 = 2.979 m from position 1 to position 2

Figure 4.3 *Section*

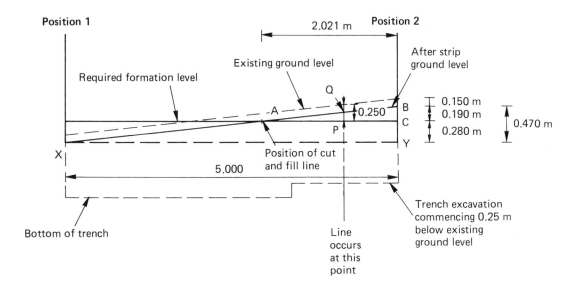

Figure 4.4 *Section*

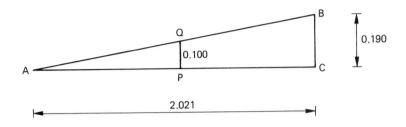

$\triangle$ ABC is similar to $\triangle$ AQP, therefore the distance AP can be found by interpolation

$$AP = \frac{0.100}{0.190} \times 2.021 = \underline{1.064 \text{ m}}$$

Figure 4.5   *Section*

Measurement

In addition to the notes in the previous chapter, the following comments also apply:

1. With regard to the oversite excavation, always measure the plan area of the building unless the slope of the site is significant.
2. The measurement of the reduce level excavation can be very complicated if taken to the extreme. If the plan shape is irregular the depths of excavation may be averaged, and Simpsons Rule applied. However, in practice, the measurement of excavation can be simplified by drawing 'give and take' lines to give regular geometric shapes; the depths of excavation are then averaged, see Figure 4.6.

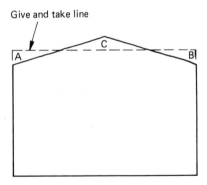

Gauged by eye the area of A + B = area of C

Figure 4.6   *Plan*

Sometimes the average depth may be calculated by a weighting method, as follows:

Example

	1	2	3	4
	1.020	1.070	1.050	1.100
A	×	×	×	×

	1.050	1.060	1.070	1.150
B	×	×	×	×

Grid of site levels

	1.100	1.100	1.110
C	×	×	×

Average

A1	1.020
A2	1.070
A3	1.050
A4	1.100
B1	1.050
B2	1.060
B3	1.070
B4	1.150
C1	1.100
C2	1.100
C3	1.110

11) 11.880

Average depth   1.080

Weighted average

Weighting	× 1	× 2	× 3	× 4
A1	1.020	A2 1.070	B3 1.070	B2 1.060
A4	1.100	A3 1.050		
B4	1.150	B1 1.050		
C1	1.100	C2 1.100		
C3	1.110			
5 =	5.480	4 = 4.270	1 = 1.070	1 = 1.060
		× 2 wgt.	× 3 wgt.	× 4 wgt.
		8 = 8.540	3 = 3.210	4 = 4.240

Summary

5 =   5.480
8 =   8.540
3 =   3.210
4 =   4.240
20) 21.470

Weighted average depth   1.074

N.B. Never average ground levels across a cut and fill line as this could give an answer of no reduce level excavation and no additional filling which is obviously incorrect.

Average ground levels either side of the cut and fill line to give an amount of reduce level excavation and an amount of additional filling.

Trench excavation

In this approach the steps in the trenches are ignored and it is thus initially assumed that the bottom of the trench is level all around the building. After calculating the centre line girths of the trench either side of the 0.25 m line (see above and SMM7 D.20.2.6.*.1), measure the trench excavation either side of the 0.25 m line in cubic metres based on depth from formation level or the average depth from after strip level which varies to the assumed level bottom of the trench. To be accurate the trench excavation should be measured piecemeal between positions of existing ground levels. After the trench has been measured to the assumed level bottom then add the excavation for the steps in slices, see Figure 4.7.

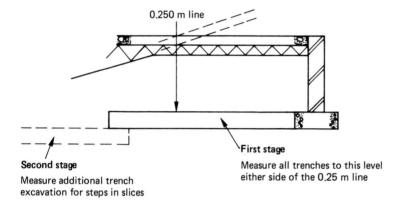

Figure 4.7   *Section*

For description purposes, the maximum depth of the trench excavation must include the steps.

Earthwork support

This follows exactly the measurement of trench excavation, i.e. measure support for both faces of the assumed level trench and add both faces for the additional excavation for the steps in slices as before. Only measure earthwork support to the faces of steps if they exceed 0.25 m high.

Concrete in foundations

Measure the concrete in the trench all round ignoring the steps and then add the additional concrete for each step – do not distinguish between the different thicknesses of concrete at the steps, see Figure 4.8.

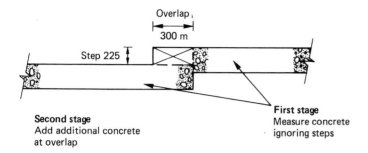

Figure 4.8   *Section through step*

Measure formwork to the vertical face of concrete at steps.

Brickwork

This follows the system of measurement for trenches, i.e. measure the brick walls all round the building from the top of concrete of the assumed level foundation to the damp course level, then add the area of brickwork in steps in slices.

Facing brickwork

The height of facing brickwork is averaged if the finished ground level varies around the building.

Filling

Measure the standard bed of filling over the whole building and measure the additional filling, including the standard bed, to the fill side of the cut and fill line and then deduct the standard bed which is over measured. The depth of hardcore over the fill area is averaged.

Flow chart for a foundation on a sloping site

SMM ref & unit		Measurement
	Start	

NIL	Calculate depths of cut and fill, and position of cut and fill line. Plot on drawing

Excavate oversite

D.20.2.1.1.0 m²	Plan area of building to external face of foundation projection

Disposal of oversite excavation

D.20.8.3.2.1 m³	Cube up above plan area by thickness of oversite excavation

Surface treatments

D.20.13.2.3.0 m²	Same area as excavate oversite

Excavate to reduce levels

D.20.2.2.2.0 m³	Volume of reduce level excavation, viz. average length of reduce level excavation measured from the cut and fill line to the external face of foundation projection × the width measured between the external faces of foundation projection × the average depth of reduce level excavation

Disposal of reduce level excavation

D.20.8.3.1.0 m³	Same volume as reduce level excavation above

Information only

NIL	Calculate position of line where trench excavation occurs 0.25 m below existing ground level. See SMM D.20.2.6.0.1. Plot on drawing, called 0.25 m line

Continued

SMM ref & unit		Measurement

Earthwork support to reduce levels

D.20.7.1.3.0 m²	Area of vertical face to sides of reduce level and oversite excavations which in total exceeds 0.25 m deep. See SMM D.20.7.M9(a), viz. length of each side is as calculated for the reduce level excavation, but measured from the 0.25 m line where applicable × the average depth of reduce level and oversite excavation of each side

Excavate trenches

D.20.2.6.2.0 m³	Volume of trench excavation ignoring steps in the foundation, viz. total of measurements to equal centre line girth of trench × width of trench × depth of trench calculated from highest level of foundation, i.e. 99.300. N.B. For description purposes only calculate maximum depth of trench including steps

Disposal of trench excavation

D.20.8.3.1.0 m³	Same volume as trench excavation above

Excavate trenches exceeding 0.25 m below existing ground level

D.20.2.6.2.1 m³	Volume of trench excavation exceeding 0.25 m below existing ground level, viz. centre line girth of trench from 0.25 m line under reduce level × width of trench × depth of trench

Adjustment of excavate trenches

D.20.2.6.2.0 m³	Deduct same volume as last item

Continued

Flow chart for a foundation on a sloping site – continued

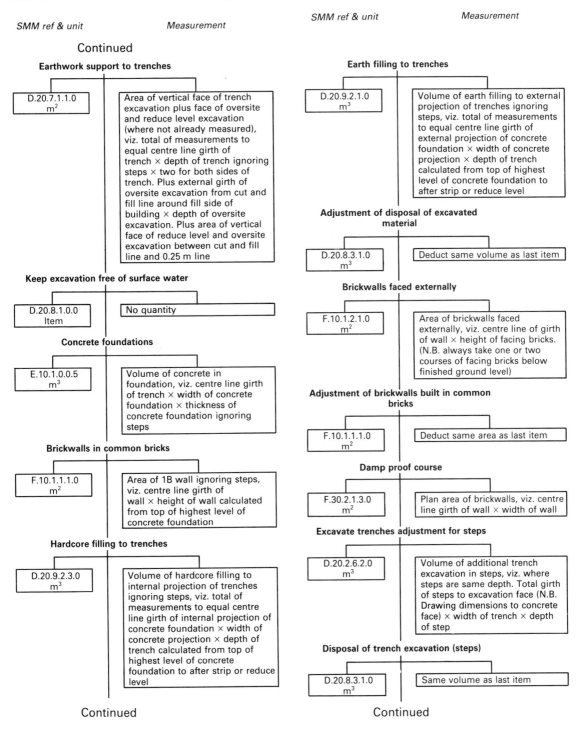

SMM ref & unit   Measurement   SMM ref & unit   Measurement

Continued

Earthwork support to trenches

D.20.7.1.1.0
m²
Area of vertical face of trench excavation plus face of oversite and reduce level excavation (where not already measured), viz. total of measurements to equal centre line girth of trench × depth of trench ignoring steps × two for both sides of trench. Plus external girth of oversite excavation from cut and fill line around fill side of building × depth of oversite excavation. Plus area of vertical face of reduce level and oversite excavation between cut and fill line and 0.25 m line

Keep excavation free of surface water

D.20.8.1.0.0
Item
No quantity

Concrete foundations

E.10.1.0.0.5
m³
Volume of concrete in foundation, viz. centre line girth of trench × width of concrete foundation × thickness of concrete foundation ignoring steps

Brickwalls in common bricks

F.10.1.1.1.0
m²
Area of 1B wall ignoring steps, viz. centre line girth of wall × height of wall calculated from top of highest level of concrete foundation

Hardcore filling to trenches

D.20.9.2.3.0
m³
Volume of hardcore filling to internal projection of trenches ignoring steps, viz. total of measurements to equal centre line girth of internal projection of concrete foundation × width of concrete projection × depth of trench calculated from top of highest level of concrete foundation to after strip or reduce level

Continued

Earth filling to trenches

D.20.9.2.1.0
m³
Volume of earth filling to external projection of trenches ignoring steps, viz. total of measurements to equal centre line girth of external projection of concrete foundation × width of concrete projection × depth of trench calculated from top of highest level of concrete foundation to after strip or reduce level

Adjustment of disposal of excavated material

D.20.8.3.1.0
m³
Deduct same volume as last item

Brickwalls faced externally

F.10.1.2.1.0
m²
Area of brickwalls faced externally, viz. centre line of girth of wall × height of facing bricks. (N.B. always take one or two courses of facing bricks below finished ground level)

Adjustment of brickwalls built in common bricks

F.10.1.1.1.0
m²
Deduct same area as last item

Damp proof course

F.30.2.1.3.0
m²
Plan area of brickwalls, viz. centre line girth of wall × width of wall

Excavate trenches adjustment for steps

D.20.2.6.2.0
m³
Volume of additional trench excavation in steps, viz. where steps are same depth. Total girth of steps to excavation face (N.B. Drawing dimensions to concrete face) × width of trench × depth of step

Disposal of trench excavation (steps)

D.20.8.3.1.0
m³
Same volume as last item

Continued

Flow chart for a foundation on a sloping site – continued

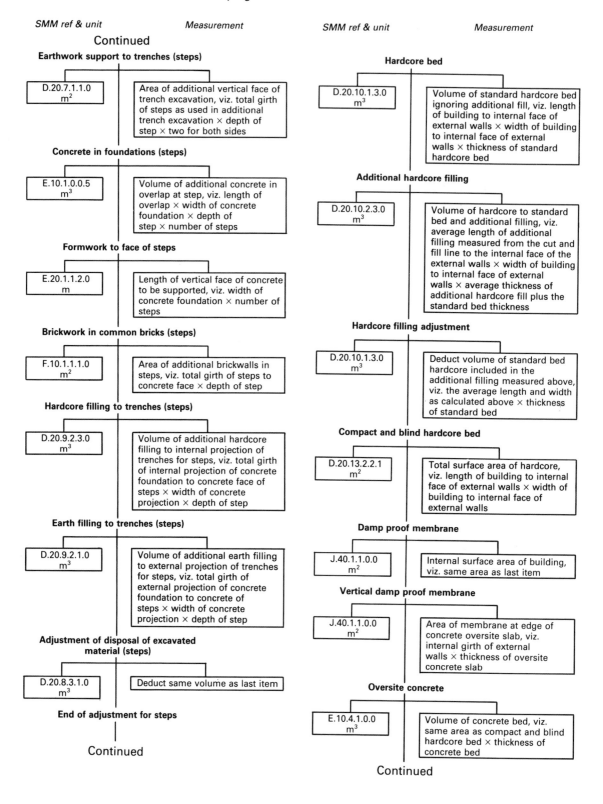

SMM ref & unit   **Measurement**

Continued

Earthwork support to trenches (steps)

| D.20.7.1.1.0 m² | Area of additional vertical face of trench excavation, viz. total girth of steps as used in additional trench excavation × depth of step × two for both sides |

Concrete in foundations (steps)

| E.10.1.0.0.5 m³ | Volume of additional concrete in overlap at step, viz. length of overlap × width of concrete foundation × depth of step × number of steps |

Formwork to face of steps

| E.20.1.1.2.0 m | Length of vertical face of concrete to be supported, viz. width of concrete foundation × number of steps |

Brickwork in common bricks (steps)

| F.10.1.1.1.0 m² | Area of additional brickwalls in steps, viz. total girth of steps to concrete face × depth of step |

Hardcore filling to trenches (steps)

| D.20.9.2.3.0 m³ | Volume of additional hardcore filling to internal projection of trenches for steps, viz. total girth of internal projection of concrete foundation to concrete face of steps × width of concrete projection × depth of step |

Earth filling to trenches (steps)

| D.20.9.2.1.0 m³ | Volume of additional earth filling to external projection of trenches for steps, viz. total girth of external projection of concrete foundation to concrete of steps × width of concrete projection × depth of step |

Adjustment of disposal of excavated material (steps)

| D.20.8.3.1.0 m³ | Deduct same volume as last item |

End of adjustment for steps

Continued

SMM ref & unit   **Measurement**

Hardcore bed

| D.20.10.1.3.0 m³ | Volume of standard hardcore bed ignoring additional fill, viz. length of building to internal face of external walls × width of building to internal face of external walls × thickness of standard hardcore bed |

Additional hardcore filling

| D.20.10.2.3.0 m³ | Volume of hardcore to standard bed and additional filling, viz. average length of additional filling measured from the cut and fill line to the internal face of the external walls × width of building to internal face of external walls × average thickness of additional hardcore fill plus the standard bed thickness |

Hardcore filling adjustment

| D.20.10.1.3.0 m³ | Deduct volume of standard bed hardcore included in the additional filling measured above, viz. the average length and width as calculated above × thickness of standard bed |

Compact and blind hardcore bed

| D.20.13.2.2.1 m² | Total surface area of hardcore, viz. length of building to internal face of external walls × width of building to internal face of external walls |

Damp proof membrane

| J.40.1.1.0.0 m² | Internal surface area of building, viz. same area as last item |

Vertical damp proof membrane

| J.40.1.1.0.0 m² | Area of membrane at edge of concrete oversite slab, viz. internal girth of external walls × thickness of oversite concrete slab |

Oversite concrete

| E.10.4.1.0.0 m³ | Volume of concrete bed, viz. same area as compact and blind hardcore bed × thickness of concrete bed |

Continued

Flow chart for a foundation on a sloping site – continued

SMM *ref & unit*   *Measurement*

Continued

Adjustment of soil to perimeter of building

| D.20.10.1.2.3 m³ | Volume of soil filling around external projection of concrete foundation of external walls (usually topsoil), viz. centre line girth of external projection of concrete foundation of external walls × width of concrete projection × thickness of filling |

Check drawings and taking-off list for outstanding items

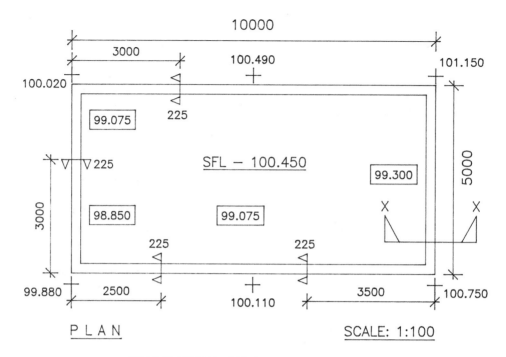

P L A N   SCALE: 1:100

FOUNDATIONS ON A SLOPING SITE

+101.150 — Existing ground levels
5 m X 5 m grid.
99.300 — Bottom of trench levels

225 — Step in bottom of trench showing
position of concrete face, direction
and depth of step — see detail below:

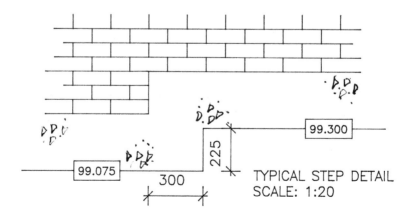

TYPICAL STEP DETAIL
SCALE: 1:20

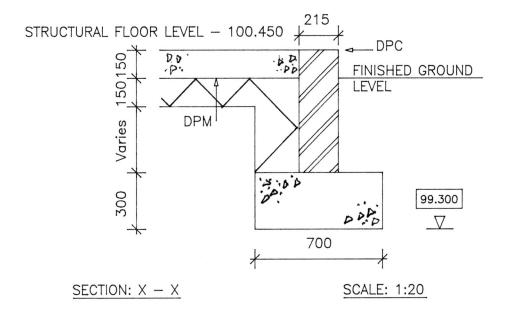

SECTION: X – X   SCALE: 1:20

SPECIFICATION

1. Topsoil: 150 mm deep – to be excavated and
   deposited in spoil heaps 30 m from excavation
2. All surplus soil taken to a tip provided by
   the contractor.
3. No ground water.
4. Concrete: 15N/20 mm
5. Brickwork a) Common bricks in cement mortar
                 (1:3) in english bond.
              b) Facing bricks to be Newdigate
                 facings in cement mortar (1:3)
                 in english bond pointing with
                 a flush joint as work proceeds.
6. DPC: Pitch polymer bedded in cement mortar
   (1:3) lapped 150 mm.
7. DPM: 1000 gauge polythene lapped 150 mm at
   all joints.
8. Filling to make–up levels – hardcore.

Drawing Number

Name of Project

date

Name 1

Sloping Site Foundation 1

Taking Off List

1. Excavate vegetable soil and disposal of soil.
2. Surface treatments
3. Reduce level excavation and disposal of soil.
4. Earthwork support to reduce level excavation
5. Excavate trenches and disposal of soil. (Standard)
6. Earthwork support to trenches
7. Keep excavations free of surface water.
8. Concrete in foundations (Standard)
9. Brickwork up to damp proof course (Standard)
10. Backfilling to trenches (Standard)
11. Facing brickwork and adjustment.
12. Damp proof course
13. Steps in foundation (adjust items 5-10 above)
14. Hardcore bed and adjustment for additional fill.
15. Blind hardcore bed
16. Damp proof membrane
17. Oversite concrete
18. Adjustment of top soil to perimeter

Page Nrs.

1	11
2	12
3	13
4	14
5	15
6	16
7	17
8	18
9	19
10	20

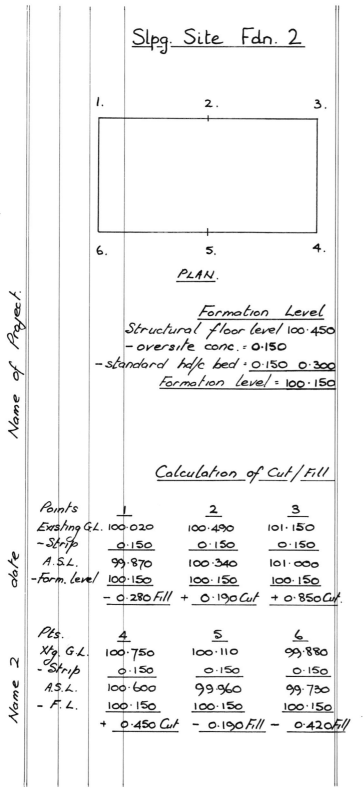

Slpg. Site Fdn. 2

1.   2.   3.

6.   5.   4.

PLAN.

Formation Level

Structural floor level 100·450
- oversite conc. = 0·150
- standard hd/c bed = 0·150 0·300
Formation level = 100·150

Calculation of Cut/Fill

Points	1	2	3
Existing G.L.	100·020	100·490	101·150
- Strip	0·150	0·150	0·150
A.S.L.	99·870	100·340	101·000
- Form. level	100·150	100·150	100·150
	− 0·280 Fill	+ 0·190 Cut	+ 0·850 Cut

Pts.	4	5	6
Xtg. G.L.	100·750	100·110	99·880
- Strip	0·150	0·150	0·150
A.S.L.	100·600	99·960	99·730
- F.L.	100·150	100·150	100·150
	+ 0·450 Cut	− 0·190 Fill	− 0·420 Fill

Name of Project.

date

Name 2

Sometimes it is convenient to draw a rough sketch in dimensions and letter or number salient points, thus making reference in dimensions easier.

This calculation ascertains the formation level which is the level of the underside of the minimum hardcore bed required by the designer, and indicates the level at which the cut and fill line occurs.

In these calculations the formation level (FL) is compared with the after strip level (ASL).
Where the FL is lower than the ASL, cutting or digging is required as at points 2, 3 and 4, and where the FL is higher than the ASL, filling is required as at points 1, 5 and 6.

Name of Project.

Name 3   date

Slpg. Site Fdn. 3

Cut and Fill Line

The position of the cut and fill line is calculated and drawn onto the foundation plan similar to a contour line. Any ground which is at a higher level than the formation level represented by the cut and fill line has to be excavated or reduced, and any ground lower will have to be filled.

Position of Cut and Fill Line

C&F Line falls between 1&2

$$
\begin{array}{ll}
\text{Point 1} & 0.280 \text{ fill} \\
\text{" } 2 & 0.190 \text{ cut} \\
\hline
& 0.470
\end{array}
$$

By interpolation
$$
\frac{0.190}{x} = \frac{0.470}{5.000}
$$
$$
x = 2.021 \text{ m}
$$

VERTICAL SECTION – POINTS 1-2

C&F Line falls between 4&5

$$
\begin{array}{ll}
\text{Point 4} & 0.450 \text{ cut} \\
\text{" } 5 & 0.190 \text{ fill} \\
\hline
& 0.640
\end{array}
$$

By interpolation
$$
\frac{0.450}{y} = \frac{0.640}{5.000}
$$
$$
y = 3.516 \text{ m}
$$

VERTICAL SECTION – POINTS 4-5

Slpg Site Fdn. 4

Cut and Fill Line (Ctd)

The depths of cut or fill and the position of the C&F line as shown here are drawn to scale on the original foundation plan.

PLAN

Excavate o/site.

Ext. dims. 10·000 × 5·000

+ Conc. proj.
br. 0·700
- wall 0·215

2/½/ 0·485 = 0·485   0·485
        10·485   5·485

10·49   Excavtg. topsoil for pres.   SMM D20.2.1.1.0
5·49   av. 150 mm dp.

&

Disp. excavtd. mat. on site in   SMM D20.8.3.2.1
sp. hps. av. 30m fr. excavtn.
Cube × 0·15 = _____ m³

Slpg. Site Fdn. 5

Surface treatments

10.49	Surf. treatments comp.
5.49	bttms. of excavtns.

SMM D20.13.2.3.0

Reduce Levelling.

C'f Line to point 2:	2.021
points 2 – 3	5.000
	7.021
C'f line to point 4 =	3.516
2)	10.537
Average	5.269
conc. proj. ½/0.485	0.243
	5.512
Trapezium dims. ×	5.485

The plan shape of the reduced levels is a trapezium. Formula for the area of a trapezium is the average length of the parallel sides times the perpendicular distance between them. See Appendix 7. This area is then multiplied by the average depth of digging to give the volume of reduced level excavation.

Av. depth of R.L	
c'f line	0.000
pt. 2	0.190
" 3	0.850
" 4	0.450
c'f line	0.000
5)	1.490
	0.298

5.51	Excavtg. to r.l. max.
5.49	depth ≤ 1.00 m
0.30	

SMM D20.2.2.2.0

&

Disp. excavtd. mat. off site.

SMM D20.8.3.1.0

Name of Project.

Name   date

Name 5

<u>Slpg. Site Fdn. 6.</u>

<u>*Position of 250 mm total*</u>
<u>*excavation depth below*</u>
<u>*existing ground level.*</u>

Total depth	0·250
− *site strip*	0·150
depth of reduce level	0·100

<u>250 *line occurs between C&F a pt 2.*</u>
By *interpolation* $\dfrac{0·100}{x} = \dfrac{0·190}{2·021}$

$x = 1·064\,m$

<u>250 *line occurs between C&F a pt. 4.*</u>
By *interpolation* $\dfrac{0·100}{Y} = \dfrac{0·450}{3·516}$

$Y = 0·781\,m$

SMM D20.2.0.0.1 requires any excavation which commences exceeding 0.25 m below existing ground level to have the commencing level stated in the description. Therefore this depth, i.e. 250 mm, must be plotted on the foundation plan. See plan on Slpg Site Fdn 4.

Excavation of vegetable soil over site of 150 mm has already been measured. Therefore it is necessary to calculate where the reduce level excavation is 100 mm deep, giving a total excavation depth of 250 mm.

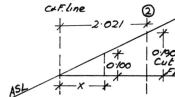

VERTICAL SECTION C&F to pt. 2.

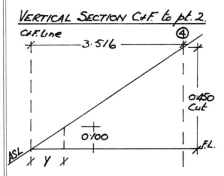

VERTICAL SECTION C&F to pt. 4.

Slpg. Site Fdn. 7

Earthwork support
to reduce levels etc.

	length	depth
a. between 250 line & pt. 2.	2·021	0·250
	− 1·064	0·340
	0·957	2)0·590
		0·295
b. between pt. 2 & pt 3	5·000	0·340
+ conc. proj. ±/0·485 = 0·243	1·000
	5·243	2)1·340
		0·670
c. between pt. 3 & pt 4	5·485	1·000
		0·600
		2)1·600
		0·800
d. between pt. 4 & 250 line	3·516	0·600
+ conc proj. ±/0·485 : 0·243	0·250
	3·759	2)0·850
	− 0·781	0·425
	2·978	

Earthwork support to reduced level and vegetable soil excavation is measured in two parts. That which is ≤0.25 m deep and is vertically above the external face of the trench is included with the trench support and that which is >0.25 m deep is measured separately. (See SMM D20.7–M9(a).)

0·96	Earthwork suppt. max.	(a
0·30	depth ≤ 1·00 m dist.	
5·24	bet. opp. faces	(b
0·67	> 4·00 m	
5·49		(c
0·80		
2·98		(d
0·43		

SMM D20.7.1.3.0

Name of Project.   date   Name 8

Slpg. Site Fdn. 8

<u>Calculation of standard and maximum trench depth</u>

	1.	2.	3
	ASL 99·870	FL 100·150	FL 100·150
Standard Trench Level	99·300	99·300	99·300
Standard Trench depth	0·570	0·850	0·850
+ step	0·225		
	0·795		

	4.	5.	6
	FL 100·150	ASL 99·960	ASL 99·730
Standard Trench Level	99·300	99·300	99·300
Standard Trench depth	0·850	0·660	0·430
+step		0·225	$\frac{2}{}$0·225 0·450
		0·885*	0·880

Maximum trench depth.

<u>Standard trench excavtd. from R.L.</u>

0·850 deep trench.   ₵ girth
cₑf.line to pt. 2.   2·021
pt.2 – pt.3   5·000
pt.3 – pt.4   5·000
pt.4 to cₑf.line   3·516
                              15·537
– passings $^{2}/\frac{4}{}$/0·215 :   0·430
₵ girth   15·107

The maximum depth of trench excavation is required for maximum depth classification. (See SMM D20.2.2 to 6.1 to 4.)

Excavation of trenches at points 2, 3 and 4 commences at formation level (FL) and has a standard depth. That at points 1, 5 and 6 starts at after strip level (ASL), and has various trench depths.
The measurement of trench excavation commencing >0.25 m below existing ground level must be measured separately (see previous note).

Trench excavation etc. is measured to the highest foundation level, i.e. 99.300 all round, and then adjusted later for steps.

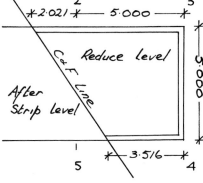

<u>Slpg.</u> <u>Site Fdn. 9</u>

<u>Trenches (Ctd)</u>

<u>Standard trench</u>
<u>excavtd. from A.S.L</u>
₵ girth
10·000

            5·000
        2/15·000
Ext. bwk. face 30·000
            4/2½/
-passings   ½/0·215· 0·860
      ₵ girth   29·140
- ₵ girth trench in R.L.15·107
      ₵ girth   14·033

                Depth
Pt. 1   0·570
C&F Line   0·850
C&F Line   0·850
Pt. 5   0·660
Pt. 6   0·430
            5)3·360
Average depth 0·672

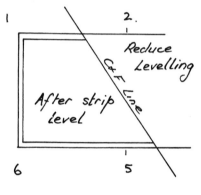

1   2.

                        Reduce
                          Levelling

After strip
  level

6   5

15·11	Excavtg. tr. >0·30m (tr. below max. depth ≤ 1·00m	SMM D20.2.6.2.0
0·70	(R.L	
0·85		
14·03	(tr. below ASL	
0·70		
0·67	+	
20·56	(steps in fdns.	
0·70		
0·23	Disp. excavtd mat. off site	SMM D20.8.3.1.0

The back filling to trench is measured later, therefore the excavated soil is carted away and not backfilled in trench as in the first example.

Slpg. Site Fdn. 10

Trenches (Ctd)

<u>Adjustment for trenches</u>
<u>commencing ≥ 0·25m bxtg. g.l.</u>

⊄ girth
girth of tr. in r.l.   15·107
– caf to 250 line
            1·064
            0·781   1·845
                            13·262

<u>Av. depth below xtg.g.l.</u>
                        0·250
pt.2. strip 0·150
       r.l. 0·190·0·340
pt.3   0·150
              0·850·1·000
pt. 4   0·150
              0·450·0·600
                    0·250
                5) 2·440
                    0·488
            <u>Say   0·500</u>

Calculation of average depth, commencing level below existing ground level.

13·26		Excavtg. tr. >0·30m max depth ≤ 1·00 m comm. at r.l. av. 0·50m below xtg. g.l.	SMM D20.2.6.2.1
0·70			
0·85			

&

<u>Ddt</u> excavtg. tr. >0·30m   SMM D20.2.6.2.0
    max depth ≤ 1·00m
    a.b.

<u>Slpg.</u> Site Fdn. 11

<u>Earthwork support.</u>
<u>to trenches</u>

2/	15.11	Earthwk. suppt. (tr. below
	0.85	max. depth ≤ 1.00m r.l.
2/	14.03	dist. bet. opp. (tr. below
	0.67	faces ≤ 2.00m ASL
	15.43	(site strip
	0.15	ASL
	1.85	(site strip
	0.20	∝ r.l.
2/	20.56	(steps
	0.23	

SMM D20.7.1.1.0

<u>site strip ASL</u>
& tr. below ASL   14.033
2/2/½/
+passings 3/0.700: 1.400
ext girth   15.433

<u>site strip ∝ r.l.</u>
bet. caf line ∝ 250 line 1.064
0.781
1.845

<u>depth</u>
strip   0.150
strip ∝ r.l.   0.250
2) 0.400
Av.   0.200

	<u>Item</u>	Disp. surf. water

SMM D20.8.1.0.0

Name of Project.

Name date

Name 11

Slpg. Site Fdn 12

Conc. fdn.

29·14	
0·70	
0·30	
2/2/ 0·30	
0·70	
0·23	

In situ conc (15 N/20 mm) fdns. poured on or against earth or unblinded hd/c   (steps

SMM E10.1.0.0.5

The foundation concrete is measured ignoring any steps.

There is no adjustment needed for filling to excavations as all soil is carted away off site. See previous note.

Brickwork

Min. hgt of bwk.

```
              SFL.  100·450
Bottom of tr. 99·300
  +conc. fdn  0·300 99·600
                     0·850
```

29·14	
0·85	
19·36	
0·23	

Bk. walls 1B thick vert. in c. bks. in Eng. bond in   (steps c.m. (1:3)

SMM F10.1.1.1.0

There is no adjustment of filling needed. See note next concrete foundation above.

Backfilling to tr.

Int. proj. girth to tr. in r.l.

```
      2/2/¢ girth ab.  15·107
-passings. bwk. 2/0·215=0·430
      3/3/
   + proj. 2/0·243 0·486 0·916
                          14·191
```

Name of Project.

Name of date

Name 12

SECTION.

(labels on section: BB: / A / 6A / GL / hardcore bed / hardcore filling to internal projection / Earthfilling to external projection / BB: / AA:)

Height of brickwork ignoring steps.

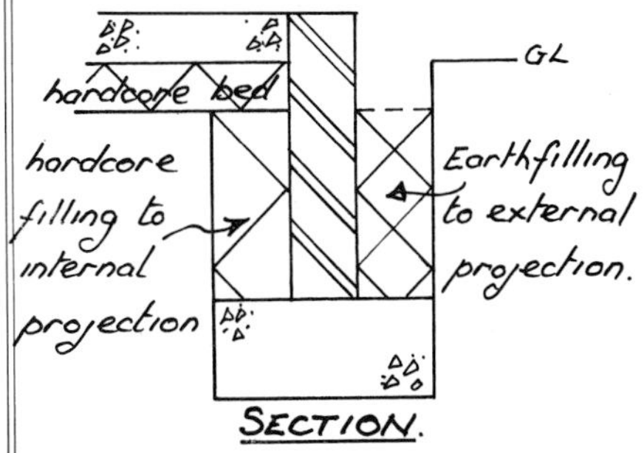

Slpg. Site Fdn. 13

Backfilling to tr. (Ctd)

$$\underline{depth}$$
tr. depth   0·850
− conc. fdn.   0·300
0·550

$$\underline{Int. proj. girth}$$
to tr. from ASL.
₵ girth ab   14·033
− passings as above   0·916
13·117

$$\underline{depth}$$
tr. depth   0·672
− conc. fdn. 0·300
0·372

The first passing adjustment adjusts the ₵ girth of trench to the inside face of brickwork and then the second passing adjustment adjusts the girth to the ₵ of the internal concrete projection.

14·19			Hd/c filling to (tr. in r.l.
0·24			excavtns. av. thick.
0·55			> 0·25 m obtained
13·12			off site, selected (tr. from
0·24			brick or stone. (ASL
0·37			
17·98			(step
0·24			
0·23			

SMM D20.9.2.3.0

₵ bwk. top step 14·070
− passings as above 0·916
13·154
₵ bwk. bttm. step 5·285
− passings
bwk. ³/₊/0·215: 0·215
proj. ³/₊/0·243: 0·243: 0·458: 4·827
₵ int. proj. 225 mm. hi. 17·981

<u>Slpg. Site Fdn. 14.</u>

<u>Backfilling to tr. (Ctd)</u>

<u>Ext. proj. girth</u>
<u>Tr. in r.l.</u>
& girth a.b. 15·107
+ passings a.b. 0·916
16·023

<u>Tr. from A.S.L</u>
& girth a.b. 14·033
+ passings a.b. 0·916
14·949

<u>Same depths as for hd/c filling</u>

16·02	<u>Ddt</u> disp. excavtd (tr. in r.l. SMM D20.8.3.1.0
0·24	mat. off site a.b.
0·55	
14·95	(tr. from
0·24	♂ (A.S.L
0·37	
20·73	(steps
0·24	<u>Add</u> filling to excavtns. SMM D20.9.2.1.0
0·23	av. thickness >0·25m
	arising fr. excavtns.

<u>steps</u>
& bwk. top step a.b. 14·070
+ passings a.b. 0·916
14·986
& bwk. bttm. step a.b. 5·285
+ passings a.b. 0·458 5·743
& ext. proj. 225 mm. hi. 20·729

<u>Facings adj.</u>
NB. The level of the grd.
around the bdg. is made
up to 150 mm below d.p.c.

If the finished and existing ground levels are the same, then the depth of facings around the building will vary.

<u>Slpg.</u> Site Fdn. 15

<u>Facing adj. (Ctd)</u>

<u>Depth</u>

Fin. g. l. to dpc. 0·150
below g.l. 0·150
0·300

29·14 0·30	Bk. walls facework one side 1B thick vert. in c. bks. & Newdigate Fcg. bks in Eng. bond in c.m (1:3) & ptg. with a flush jt. a.w.p.

SMM F10.1.2.1.0

&

	<u>Ddt</u> 1B walls in c. bks. a.b.

SMM F10.1.1.1.0

<u>D.p.c.</u>

29·14 0·22	Pitch Polymer single layer d.p.c. width ≤ 225 mm horiz. bedded in c.m.(1:3) lapped 150mm at all passings (Meas. nett)

SMM F30.2.1.3.0

Name of Project

date

Name 15

Name

Slpg. Site Fdn. 16

Adj. for steps.

All dimensions are measured from external corners of the building to the face of the concrete step.

225 mm hi. steps

1st. step.	bet. pts 1&2	3·000	
	'' '' 1&6	5·000	
bet. pts 4&6		10·000	
	−	3·500 :	6·500
			14·500
−passings	2/3/½/0·215 :		0·430
bwk. & 1st step			14·070
2nd step	bet. pts 1&6	3·000	
	'' '' 5&6	2·500	
		5·500	
−passing	1/2/½/0·215		0·215
bwk. & 2nd step		5·285 :	5·285
bwk. & all steps			19·355

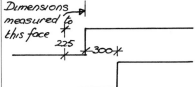

SECTION THROUGH STEP

Additional work caused by steps in foundations is measured in 225 mm high strips.

& tr. excavtn

bwk. &			19·355
+ conc. overlap	2/2/0·300:	1·200	
			20·555

Measurement of steps added back to slpg. site fdn. 9 et. seq. except:

Fmwk to face
of steps

2/2/0·70

Fmwk. for in situ conc. sides of fdns. plain vert. hgt. ≤ 250 mm

SMM E20.1.1.2.0

NB. Earthwork support is not measureable to earth face of step (Ctd)

(Ctd.) as height of step does not exceed 250mm see SMM D20·7−M9(a)

Slpg. Site Fdn. 17

Hardcore bed

	10.000	×	5.000
−bwk. 2/0.215	0.430		0.430
int. dims	9.570		4.570

9.57
4.57
0.15

Hd/c filling to make up levels av. thickness ≤ 0.25m obtained off site

The standard hardcore bed is measured overall and then adjusted for the additional filling.

SMM D20.10.1.3.0

Additional hd/c filling adjustment.

pt 1. to C&F Line	5.000
−	2.021
	2.979
−wall	0.215
	2.764
pt 6 to C&F Line	10.000
−	3.516
	6.484
−wall	0.215 6.269
	2)9.033
	Av = 4.517

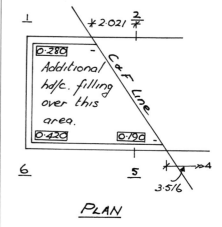

PLAN

The total thickness of hardcore including the standard bed is averaged over the trapezium-shaped area shown above and then the standard bed is adjusted.

Thickness

Pt. 1.	0.280
C&F Line	0.000
" "	0.000
Pt. 5.	0.190
Pt. 6	0.420
	5)0.890
	0.178
+ Standard bed	0.150
Total av. thickness	0.328

Slpg. Site Fdn. 18

Hardcore bed (Ctd)

4·52	Hd/c filling to make up	SMM D20.10.2.3.0
4·57	levels av. thickness	
0·33	> 0·25 m obtained	
	off site	

Standard bed adj.

4·52	Ddt ditto av thickness	SMM D20.10.1.3.0
4·57	≤ 0·25m ditto	
0·15		

9·57	Surf. treatments comp.	SMM D20.13.2.2.1
4·57	filling, blinding with	
	sand.	

&

	Flexible sheet tanking	SMM J40.1.1.0.0
	d.p.m. horiz. 1000 g.	
	polythene laid on	
	blinded hd/c	

Name of Project

Name date 18 Name

Slpg. Site Fdn 19

D.p.m. (Ctd).

Int. girth
9·570
4·570
2/14·140
28·280

	28·28		Flexible sheet d.p.m. vert. 1000 g. polythene set against bwk.	SMM J40.1.1.0.0
	0·15			

The d.p.m. must be continuous with d.p.c.

Conc. bed.

9·57	In situ conc (15N/20mm) beds, thickness ≤ 150 mm	SMM E10.4.1.0.0
4·57		
0·15		

Top soil fill to perimeter
Excavate oversite dims. 10·485
5·485
2/15·970
Ext. girth : 31·940
− Passings 4/2/1/0.243 : 0·972
& girth of proj. 30·968

30·97	Filling to make up levels av. thickness ≤ 0·25m obtained from on site spoil heaps av. 30m from excavtn.	SMM D20.10.1.2.3
0·24		
0·15		

Name of Project

date

Name 19

5

Basements

Introduction

The notes made in the previous chapters apply to this section of work with the following additional comments.

Approach

If a building comprises part basement and part strip foundation, measure the basement first followed by the strip foundation, making any necessary adjustments. If a basement is of reinforced concrete construction then the engineer may require blinding layers on which to set out the reinforcement and these may not be shown on the drawing. The taker-off should check to see if they are required and if formwork is necessary to every face, before measurement commences.

Measurement

The plan dimension for basement excavation is taken to the largest part of the basement construction, irrespective of the fact that working space may be required in addition to these dimensions. Any trenches at the bottom of the basement are described as excavating trenches but with their commencing levels stated.

Ground water

Ground water is water present in the subsoil which percolates through the sides of excavations. The ground water level, date when established and position of trial pit or bore hole must be included in the contract documentation. This level is defined as the 'pre-contract water level', see SMM D20.P1(A). Measurements in the bills of quantities are based on this level.

Changes in weather conditions during the contract period may cause the ground water level to rise or fall. It must therefore be re-established at the time each excavation is carried out, when it is defined as the 'post-contract water level', see SMM D20.P1(B).

If the pre- and post-contract ground water levels differ then the quantity of excavation below ground water level will be remeasured. It is up to the contractor to decide how to dispose of the ground water and price accordingly.

Excavating below ground water level

Only the volume of excavation occurring below the ground water level is measured and described as 'extra over any type of excavation for excavation below ground water level', see SMM D20.3.1. If the post-water level varies from that used to produce the bills of quantities then this must be remeasured, see SMM D20.3/M5.

Working space

Working space is measured where the face of the excavation is less than 600 mm from the face of formwork, rendering, tanking or protective walls. It is measured in square metres and calculated by multiplying the girth of formwork, etc. by the depth of excavation below the commencing level, see Figures 5.1 and 5.2. The item includes additional earthwork support, disposal, backfilling, etc., see SMM D20.6.M7, M8 and C2.

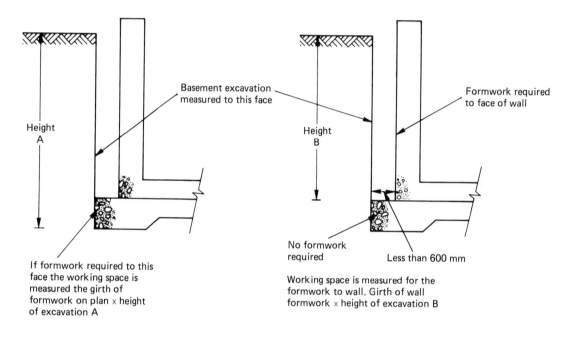

Basement excavation measured to this face

Formwork required to face of wall

Height A

Height B

No formwork required

Less than 600 mm

If formwork required to this face the working space is measured the girth of formwork on plan × height of excavation A

Working space is measured for the formwork to wall. Girth of wall formwork × height of excavation B

Figure 5.1   *Section*

Figure 5.2   *Section*

Earthwork support

Earthwork support is measured to the same dimensions as those used for the basement excavation, irrespective of the fact that working space may be required, see Figure 5.3. Any additional support caused by the working space is deemed to be included in the cost of the working space, see SMM D20.6.–C2.

If any part of the earthwork support extends below the ground water level then all the earthwork support must so be described. It is only measured when there is a corresponding item of extra over for excavating below the ground water level, see SMM D20–M10 and M11.

Earthwork support to basement trench excavation where the outside face coincides with the face of the basement excavation and does not exceed 0.25 m high is, in practice, measured and included in the basement earthwork support. There is no need to measure earthwork support to the inside face of the basement trench if it does not exceed 0.25 m high, see SMM D20.7–M9A. Earthwork support for basement trenches exceeding 0.25 m high will be measured to both faces and classified according to its distance between opposing faces.

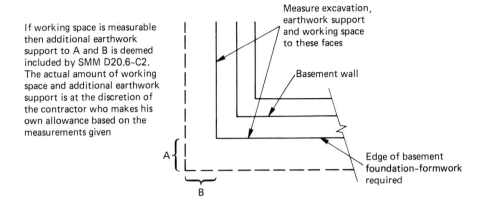

Figure 5.3 *Plan at corner*

In-situ concrete

In-situ concrete to beds, slabs, walls, etc., is measured in cubic metres and the thickness range is included in the description. If a bed is thickened under a wall then the volume of this thickening is calculated and added to the volume of the bed. The thickness range of the slab is not altered even if it goes in to the next range, see SMM E10.4–D3.

Formwork for sloping *in-situ* concrete

In-situ concrete is initially a fluid-like material and if laid to a slope exceeding 15 degrees from the horizontal, it tends to flow downhill. Therefore, formwork to the top surface of the concrete must be measured, see SMM E20.11–M7.

Flow chart for concrete basement

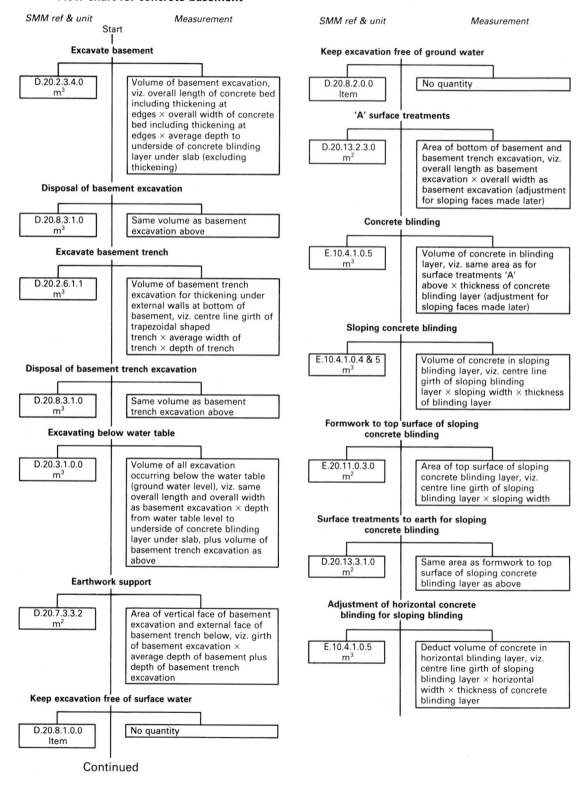

SMM ref & unit — Start — **Measurement**

Excavate basement

D.20.2.3.4.0 m³ | Volume of basement excavation, viz. overall length of concrete bed including thickening at edges × overall width of concrete bed including thickening at edges × average depth to underside of concrete blinding layer under slab (excluding thickening)

Disposal of basement excavation

D.20.8.3.1.0 m³ | Same volume as basement excavation above

Excavate basement trench

D.20.2.6.1.1 m³ | Volume of basement trench excavation for thickening under external walls at bottom of basement, viz. centre line girth of trapezoidal shaped trench × average width of trench × depth of trench

Disposal of basement trench excavation

D.20.8.3.1.0 m³ | Same volume as basement trench excavation above

Excavating below water table

D.20.3.1.0.0 m³ | Volume of all excavation occurring below the water table (ground water level), viz. same overall length and overall width as basement excavation × depth from water table level to underside of concrete blinding layer under slab, plus volume of basement trench excavation as above

Earthwork support

D.20.7.3.3.2 m² | Area of vertical face of basement excavation and external face of basement trench below, viz. girth of basement excavation × average depth of basement plus depth of basement trench excavation

Keep excavation free of surface water

D.20.8.1.0.0 Item | No quantity

Continued

Keep excavation free of ground water

D.20.8.2.0.0 Item | No quantity

'A' surface treatments

D.20.13.2.3.0 m² | Area of bottom of basement and basement trench excavation, viz. overall length as basement excavation × overall width as basement excavation (adjustment for sloping faces made later)

Concrete blinding

E.10.4.1.0.5 m³ | Volume of concrete in blinding layer, viz. same area as for surface treatments 'A' above × thickness of concrete blinding layer (adjustment for sloping faces made later)

Sloping concrete blinding

E.10.4.1.0.4 & 5 m³ | Volume of concrete in sloping blinding layer, viz. centre line girth of sloping blinding layer × sloping width × thickness of blinding layer

Formwork to top surface of sloping concrete blinding

E.20.11.0.3.0 m² | Area of top surface of sloping concrete blinding layer, viz. centre line girth of sloping blinding layer × sloping width

Surface treatments to earth for sloping concrete blinding

D.20.13.3.1.0 m² | Same area as formwork to top surface of sloping concrete blinding layer as above

Adjustment of horizontal concrete blinding for sloping blinding

E.10.4.1.0.5 m³ | Deduct volume of concrete in horizontal blinding layer, viz. centre line girth of sloping blinding layer × horizontal width × thickness of concrete blinding layer

Flow chart for concrete basement – continued

SMM ref & unit	Measurement	SMM ref & unit	Measurement

Continued

Adjustment of surface treatments to earth for sloping concrete blinding

D.20.13.2.3.0 m²	Deduct area of surface treatments compacting to bottoms of excavations, viz. centre line girth of sloping blinding layer × horizontal width of sloping blinding layer

Concrete bed and thickening

E.10.4.1.0.1 m³	Volume of concrete in bed and thickening, viz. same area as for surface treatments 'A' above × thickness of concrete bed, plus centre line girth of trapezoidal shaped thickening × average width of thickening × depth of thickening

Finish to concrete bed

E.41.3.0.0.0 m²	Area of concrete bed, viz. same area as for surface treatments 'A' above

Formwork to edge of concrete bed and thickening

E.20.2.1.3.0 m	Girth of external edge of concrete bed, viz. same girth of basement excavation as used for earthwork support

Working space

D.20.6.1.0.0 m²	Area of vertical face of excavation, viz. same area as used for earthwork support

Reinforcement

To take note from engineer's schedules	Not measured in this example

Horizontal asphalt tanking

J.20.1.4.1.1 m²	Area of horizontal asphalt tanking, viz. length to external face of brick protection walls × width to external face of brick protection walls

Continued

Concrete floor

E.10.4.2.0.1 m³	Volume of concrete in floor, viz. length to external face of concrete walls × width to external face of concrete walls × thickness of concrete floor

Concrete walls

E.10.7.2.0.1 m³	Volume of concrete in walls and attached columns, viz. centre line girth of wall × thickness of wall × height of wall from top of floor to soffit of suspended slab, plus width of column × depth of column × height as for walls × 2

Concrete suspended floor

E.10.5.1.0.1 m³	Volume of concrete in suspended floor and attached beam, viz. length to internal face of half brick superstructure walls × width to internal face of half brick superstructure walls × thickness of floor slab, plus length of beam in between piers × width of beam × depth of beam

Worked finishes to concrete floor

E.41.3.0.0.0 m²	Internal area of basement floor less wants caused by attached columns

Wall kickers

E.20.22.0.0.0 m	Length of kicker, viz. centre line girth of wall as used for concrete wall

Formwork to walls

E.20.12.0.1.1 m²	Area of both faces of concrete walls measured over columns, viz. external girth of concrete walls × height from bottom of basement floor to top of wall, plus internal girth of concrete walls × internal height of basement

Continued

Flow chart for concrete basement – continued

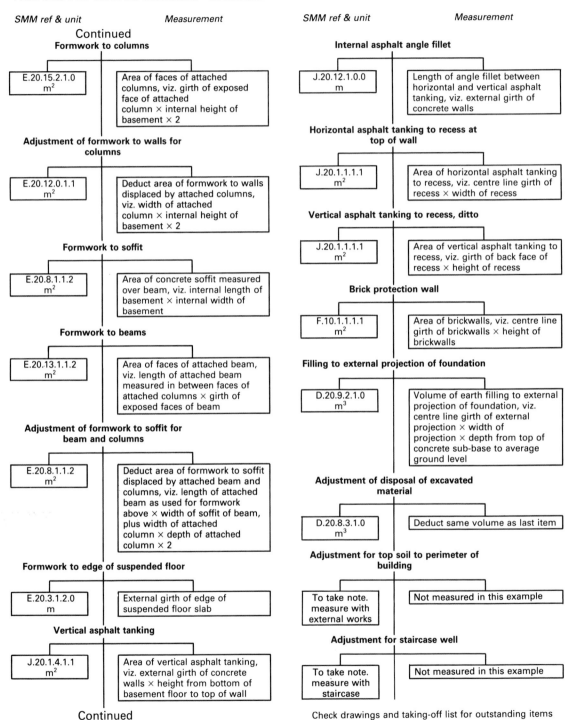

SMM ref & unit   Measurement

Continued
Formwork to columns

| E.20.15.2.1.0 m² | Area of faces of attached columns, viz. girth of exposed face of attached column × internal height of basement × 2 |

Adjustment of formwork to walls for columns

| E.20.12.0.1.1 m² | Deduct area of formwork to walls displaced by attached columns, viz. width of attached column × internal height of basement × 2 |

Formwork to soffit

| E.20.8.1.1.2 m² | Area of concrete soffit measured over beam, viz. internal length of basement × internal width of basement |

Formwork to beams

| E.20.13.1.1.2 m² | Area of faces of attached beam, viz. length of attached beam measured in between faces of attached columns × girth of exposed faces of beam |

Adjustment of formwork to soffit for beam and columns

| E.20.8.1.1.2 m² | Deduct area of formwork to soffit displaced by attached beam and columns, viz. length of attached beam as used for formwork above × width of soffit of beam, plus width of attached column × depth of attached column × 2 |

Formwork to edge of suspended floor

| E.20.3.1.2.0 m | External girth of edge of suspended floor slab |

Vertical asphalt tanking

| J.20.1.4.1.1 m² | Area of vertical asphalt tanking, viz. external girth of concrete walls × height from bottom of basement floor to top of wall |

Continued

SMM ref & unit   Measurement

Internal asphalt angle fillet

| J.20.12.1.0.0 m | Length of angle fillet between horizontal and vertical asphalt tanking, viz. external girth of concrete walls |

Horizontal asphalt tanking to recess at top of wall

| J.20.1.1.1.1 m² | Area of horizontal asphalt tanking to recess, viz. centre line girth of recess × width of recess |

Vertical asphalt tanking to recess, ditto

| J.20.1.1.1.1 m² | Area of vertical asphalt tanking to recess, viz. girth of back face of recess × height of recess |

Brick protection wall

| F.10.1.1.1.1 m² | Area of brickwalls, viz. centre line girth of brickwalls × height of brickwalls |

Filling to external projection of foundation

| D.20.9.2.1.0 m³ | Volume of earth filling to external projection of foundation, viz. centre line girth of external projection × width of projection × depth from top of concrete sub-base to average ground level |

Adjustment of disposal of excavated material

| D.20.8.3.1.0 m³ | Deduct same volume as last item |

Adjustment for top soil to perimeter of building

| To take note. measure with external works | Not measured in this example |

Adjustment for staircase well

| To take note. measure with staircase | Not measured in this example |

Check drawings and taking-off list for outstanding items

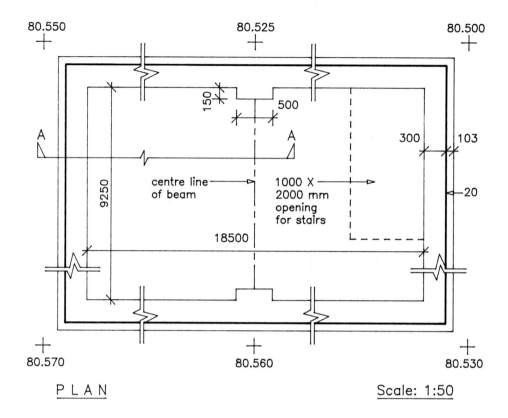

80.550   80.525   80.500

150   500

A   A   300   103

centre line of beam   1000 X
                             2000 mm   ←20
9250   opening
                             for stairs

18500

80.570   80.560   80.530

P L A N   Scale: 1:50

SPECIFICATION

1. Water table — 78.750
2. All concrete — 20N/20 mm
3. Sulphate resisting cement required
   where in contact with ground and
   below water table.
4. Basement surfaces to receive
   finishings (do not measure).
5. Brickwork — Class B Engineering
   in cement mortar (1:3).
6. Asphalt — horizontal 30 mm 3 coats
            — vertical 20 mm 2 coats
7. Do not measure reinforcement.
8. No topsoil present.

CONCRETE BASEMENT

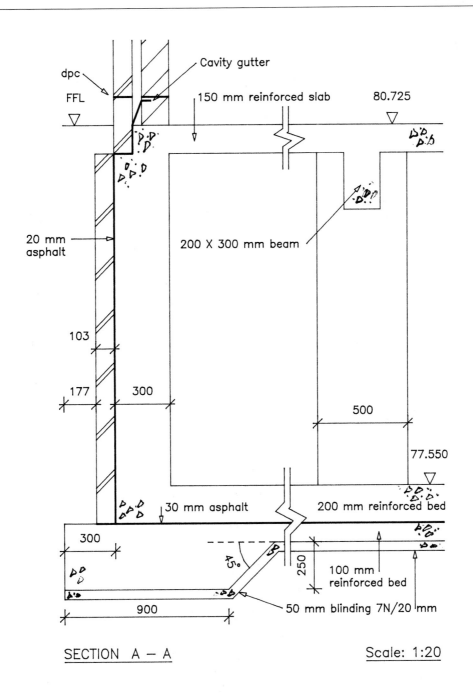

dpc

Cavity gutter

FFL

150 mm reinforced slab

80.725

20 mm asphalt

200 X 300 mm beam

103

177

300

500

77.550

30 mm asphalt

200 mm reinforced bed

300

45°

250

100 mm reinforced bed

900

50 mm blinding 7N/20 mm

SECTION A — A

Scale: 1:20

CONCRETE BASEMENT

Drawing Number.   Name of Project   date   Name 1

Concrete Basement 1

Taking Off List

1. Excavate basement and disposal of soil.
2.   "   "   trench and ditto.
3.   "   below water table.
4. Earthwork support.
5. Keep excavation free of water.
6. Surface treatments.
7. Concrete blinding layer
8. Formwork to top surface of splayed concrete.
9. Concrete in bed and toes.
10. Formwork to edge of bed.
11. Reinforcement.
12. Working space.
13. Horizontal asphalt tanking
14. Concrete in bed.
15.   "   " walls including columns.
16.   "   " slab   " beams
17. Surface treatment to concrete bed.
18. Kickers for walls.
19. Formwork to walls and columns including adj.
20.   "   " slab " beams   "
21. Vertical asphalt tanking
22. Asphalt angle fillet.
23. Brick walls.
24. Adjustment of backfilling to perimeter.

	Page Nrs.
1	11
2	12
3	13
4	14
5	15
6	16
7	17
8	18
9	19
10	20

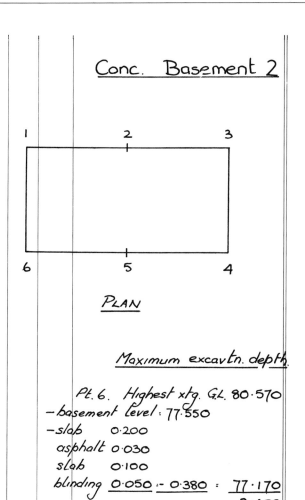

Name of Project.

date

Name 2

Conc. Basement 2

PLAN

Maximum excavtn. depth.

Pt. 6. Highest xtg. GL. 80·570
— basement level : 77·550
— slab   0·200
  asphalt 0·030
  slab   0·100
  blinding 0·050 :- 0·380 :   77·170
                            3·400
Maximum excavtn depth ≤ 4·00m.

On a sloping site the maximum
depth of excavation must be
calculated from the highest
ground level.

Ar. G. L
Pt. 1. :   80·550
Pt. 2.   ²/80·525·161·050
Pt 3 :   80·500
·· 4 :   80·530
Pt. 5.   ²/80·560 : 161·120
Pt. 6 :   80·570
         8) 644·320
Ar. g. L.   80·540

The positions of the existing
ground levels are based on a
square grid, enabling the
average ground level to be
calculated using a weighted
system.

Conc. Basement 3

Excavtn. depth

Av. g. l.   80·540
- Formation l.   77·170
            3·370

Overall basement dims.

              18·500 × 9·250
conc. wall: 0·300
proj : 0·300
²⁄₀·₆₀₀·₁·₂₀₀   1·200
       19·700   10·450

19·70	
10·45	Excavtg. basements and
3·37	the like max. depth
	≤ 4·00 m. d/p.

The dimensions for excavation
etc. are taken to the extremes
of the basement construction
as shown on the drawing, even
though additional space is
required to waterproof and to
build the half brick wall from
the outside of the basement.
See later measurement of
working space.

SMM D20.2.3.4.0

&

Disposal excavtd. mat.
off site.

SMM D20.8.3.1.0

Toe fdn.

Meas. as trapezium - width
              bottom 0·900
top.   0·900
45° splay: 0·250. 1·150
              2)2·050
     Av. width : 1·025

<u>Conc. Basement 4</u>

<u>Toe fdn. (Ctd)</u>
<u>   & girth</u>
excavtn. dims. 19·700
           10·450
       $\frac{2}{}$/30·150
         60·300
−passings $4/2/\frac{1}{2}/$1·025 = 4·100
        56·200

56·20	Excavtg. tr. width > 0·30m	SMM D20.2.6.1.1
1·03	max depth ≤ 0·25 m	
0·25	comm. av. 3·37m below	
	xtg. g.l.	

            &d;

	Disp. excavtd. mat. off site.	SMM D20.8.3.1.0

<u>Excavtg. below water table</u>
      water level 78·750
  − excavtn. depth 77·170
         1·580

19·70	E.o. any type of (basemt.	SMM D20.3.1.0.0
10·45	excavtn. for excavtg.	If the post-contract water
1·58	below grd. water level	level differs from the
56·20	(toe.	pre-contract water level
1·03		stated in the contract
0·25		documents then this
		measurement will be adjusted.
		(See D20–M5.)

Conc. Basement 5.

Earthwork suppt.
basement 3·370
toe.   0·250
           3·620

60·30	Earthwk. suppt. max depth ≤ 4·00m dist. bet. opp. faces > 4·00 m. below grd. water level
3·62	

SMM D20.7.3.3.2

Earthwork support to the external face of the basement trench which is less than 0.25 m high, but is in the same plane as the basement excavation, is included with the support to the basement. Earthwork support to the sloping side of the trench is not measureable, i.e. trench <0.25 m deep and the face slopes ≤45°. (See SMM D20.7–M9.)

Disposal of water

Item	Disposal of surface water

SMM D20.8.1.0.0

Item	Disposal of ground water

SMM D20.8.2.0.0

The item to dispose of ground water is required because an item for excavating below ground water level has been measured (see SMM D20.8.2–M12) and may be adjusted if the ground water level varies. See previous note.

Name of Project.

date

Name 5

Name of

Conc.   Basement 6

Surf. treatments

19·70 10·45	Surf. treatments comp. bttms. of excavtns.

SMM D20.13.2.3.0

Trimming slopes adj.
meas. later.

Conc. blinding layer

19·70 10·45 0·05	Sulphate resistant in situ conc (7N/20mm) beds thickness ≤ 150mm poured on or against earth

SMM E10.4.1.0.5

Slpg. conc. blinding layer

$$\underline{\text{℄ girth}}$$

ext. gi. a.b.   60·300

-passings 2/4/2½/0·900   7·200
                                           53·100
-slope passing 4/2½/0·250 1·000
                               ℄   52·100

Toe fdn.

SECTION

As the slope is 45°, the
width on plan is the same
dimension as the depth, i.e. 0.25 m.

$$\underline{\text{Width}}$$
$$= \sqrt{0·250^2 + 0·250^2}$$
$$= \underline{0·354}$$

Name 6   date   Name of Project.

<u>Conc. Basement 7.</u>

<u>Slpg. conc. blinding layer (Ctd)</u>

52·10 0·35 0·05	S.r. in situ conc (7N/20mm) beds thickness ≤ 150mm sloping > 15° poured on or against earth.

SMM E10.4.1.0.4 & 5

52·10 0·35	Fmwk. to in situ conc. top formwork slg. > 15°

SMM E20.11.0.3.0

Where concrete is poured to
slopes > 15° from horizontal,
formwork must be measured to
the top surface. (See SMM E20.11–M7

&

Surf. treatments trimming
slpg. surf. of excavtn.

SMM D20.13.3.1.0

Trimming to slopes is measured
where the slope > 15° from
horizontal. (See SMM D20.13–M20.)

52·10 0·25 0·05	Ddt. s.r. in situ conc (7N/20mm) beds horiz. a.b. ≤ 150mm thickness

SMM E10.4.1.0.5

52·10 0·25	Ddt surf. treatments bttms. of excavtns. a.b.

SMM D20.13.2.3.0

Name 7   date   Name of Project.

<u>Conc. Basement 8</u>

<u>Conc. bed.</u>

19·70	
10·45	
0·10	
56·20	
1·03	
0·25	

S.r. in situ concrete (20N/20mm) beds thickness ≤ 150 mm reinfcd.

(toe fdn.

SMM E10.4.1.0.1

The concrete thicknessing of the bed under the external wall is added to the concrete in beds and described as beds. (See SMM E20–D3C.) The thickness range stated in the description is that of the bed and is not adjusted where the thicknessing occurs. (See SMM E20–M2.)

<u>Finish to conc.</u>

19·70
10·45

Trowelling surf. of conc.

SMM E41.3.0.0.0

<u>Fmwk. to external face of bed & fdn.</u>

60·30

Fmwk. for in situ conc. sides of grd. beams & edges of beds, plain vert. hgt. 250 – 500 mm.

SMM E20.2.1.3.0

Generally where concrete is reinforced and has been designed, formwork should be measured.

Name of Project.

Name 8   date

Conc.   Basement 9

Working Space

60.30	Working space allowance
3.62	to excavtns. red. levels,
	basemts. & the like.

SMM D20.6.1.0.0

As formwork has been measured
to the external face of
concrete bed and toe
foundation and excavation has
been measured the exact size
of the basement, working space
will be required. It is
measured in square metres,
viz. the girth of formwork by
the total depth of excavation.
This item includes all
backfilling etc. (See
SMM D20.6–M7 and C2.)

To Take
All reinforcement from
Engineers   schedules

If information is not
available then a 'To Take Note'
is written into the dimensions
to ensure that items are not
forgotten.

This note will be carried
forward through the working-
up process and only crossed
through when the quantities of
steel reinforcement have been
obtained and inserted in the
bill or taking-off.

Horiz. Asphalt
18.500   ×   9.250

conc. wall	0.300		
vert. asp.	0.020		
bk. wall	0.103		
2/ 0.423	0.846		0.846
	19.346		10.096

Name 10   date   Name of Project.

Conc. Basement 10

<u>Horiz. Asp (Ctd)</u>

|19·35|
|10·10|

3 Ct. mastic asp. tanking
& damp proofing to
B.S. 1097 Col B, width
> 300 mm. horiz. 30mm
thick on conc., fin. with
a wood float, sub.
covered

SMM J20.1.4.1.1

Conc. Bed.

int. dims. 18·500 × 9·250
walls ²/0·300: 0·600   0·600
          19·100   9·850

<u>In situ concrete (20N/20mm)</u>
vibrated

|19·10|
|9·85|
|0·20|

Conc. beds thickness
150 - 450 mm. rfcd.

In the previous examples the
taking-off has more or less
followed the method of
construction, i.e. formwork for
walls followed by the concrete.
Sometimes, as is done here, it
is more convenient to measure
one trade, i.e. measure all
concrete work followed by all
formwork etc. This allows the
taker-off to use headings,
ditto etc., thus saving his
time in repeating descriptions
and the worker-up's time also.

This heading is written in the
dimensions to save repeating
the requirements of SMM E10–S1
to S5 in every description.

SMM E10.4.2.0.1

Conc. Basement. 11

In situ conc. (20N/20mm)(Ctd)

	Wall
Grd. f.l.	80·725
− slab	0·150
	80·575
−basemt. s.l.	77·550
Wall hgt.	3·025

$$\underline{\underline{\mathbb{C}\ girth}}$$
18· 500
+   9· 250
2/ 27·750
int. girth   55·500
+passings 4/2/½/0·300: 1·200
$\mathbb{C}$ 56·700

Demarcation lines of concrete measurement more or less follow the actual construction sequence on site, e.g. bed cast to external face of walls, walls cast including attached columns etc.

56·70	Conc walls thickness
0·30	150 − 450 rfcd.
3·03	
2/0·50	(attached cols.
0·15	
3·03	

SMM E10.7.2.0.1

The concrete for the attached column is included with the concrete for the wall. (See SMM E10–D6 and previous note.)

Suspd. Slab

bed	19·100	×	9·850
− wall ²/6·103	0·206		0·206
	18·894		9·644

Attached Beam

9·250
− cols ²/0·150: 0·300
8·950

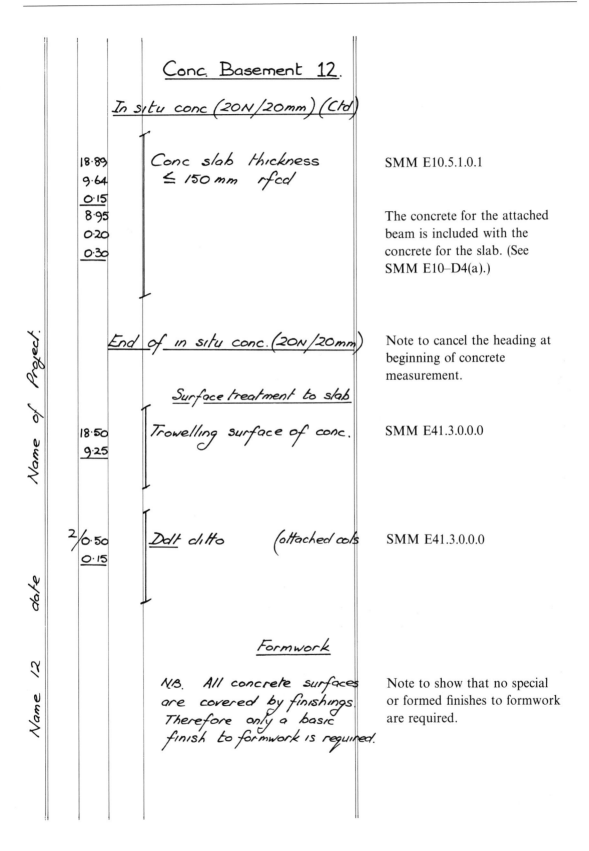

Conc. Basement 12.

In situ conc (20N/20mm) (Ctd)

18·89	
9·64	Conc slab thickness
0·15	≤ 150 mm rfcd
8·95	
0·20	
0·30	

SMM E10.5.1.0.1

The concrete for the attached beam is included with the concrete for the slab. (See SMM E10–D4(a).)

End of in situ conc. (20N/20mm)

Note to cancel the heading at beginning of concrete measurement.

Surface treatment to slab

18·50	_Trowelling_ surface of conc.
9·25	

SMM E41.3.0.0.0

2/0·50	_Ddt ditto_ (attached cols
0·15	

SMM E41.3.0.0.0

Formwork

N.B. All concrete surfaces are covered by finishings. Therefore only a basic finish to formwork is required.

Note to show that no special or formed finishes to formwork are required.

Name 12   date   Name of Project.

<u>Conc. Basement 13</u>

<u>Fmwk. (Ctd)</u>
<u>Walls.</u>

<u>56.70</u>	Fmwk. for suspd. wall kickers

SMM E.20.22.0.0.0

Formwork for wall kickers is measured along the centre line of the wall and includes both sides. (See SMM E10–M15.)

<u>External wall face</u>
4/2/½/ ₡ wall 56.700
+passings ²/0.300: 1.200
ext. girth <u>57.900</u>

<u>Height</u>
conc. wall 3.025
+ " bed 0.200
ext height <u>3.225</u>

<u>Internal wall face</u>
4/2/½/ ₡ wall 56.700
−passings ²/0.300: 1.200
int. girth <u>55.500</u>

57.90	Fmwk. for in situ.conc. (ext.
3.23	walls, vert. hgt. (face
55.50	>3.00m above (int.
3.03	floor level (face

SMM E20.12.0.1.1

Where the height of the wall exceeds 3.00 m above floor level, all of the area of formwork to the wall is described as such. (See SMM E10–D9.)

There is no deduction of formwork for the wall kicker. (See SMM E10–M10.)

<u>Girth of col.</u>
face 0.500
returns ²/0.150:0.300
<u>0.800</u>

(left margin, rotated) Name of Project.   date   Name 13

Conc Basement 14

<u>Fmwk (Ctd)</u>
<u>Columns</u>.

| 2/ | 0·80 | | Fmwk. for in situ conc | SMM E20.15.2.1.0 |
| | 3·03 | | cols. attached to walls | |

reg. shape rectangular.
hgt. > 3·00m above flr. L.
(<u>In Nr. 2</u>)

Additional information can be
given if deemed necessary. The
height is given as for walls.

It is a requirement of SMM
that the number of columns is
stated.

<u>Adj. of fmwk. to walls</u>

| 2/ | 0·50 | | <u>Ddt</u> fmwk. for walls | SMM E20.12.0.1.1 |
| | 3·03 | | a.b. | |

<u>Soffit</u>

| 18·50 | | Fmwk. for in situ conc. | SMM E20.8.1.1.2 |
| 9·25 | | soffits of slabs thickness | |

≤ 200 mm, horiz. hgt.
to soff. ≤ 4·50m.

Girth of Beam
soff.   0·200
sides 2/0·300·0·600
              0·800

| 8·95 | | Ditto beams attached to | SMM E20.13.1.1.2 |
| 0·80 | | slabs reg. shape rect. | |

hgt. to soff. ≥ 4·50 m
(<u>In Nr. 1</u>)

Name of Project.

Name date

Name 14

Conc. Basement 15.

Beam (Ctd)

	8·95
2/	0·20
	0·50
	0·15

Ddt fmwk. to soff. (beam of slabs a.b.

(cols

SMM E20.8.1.1.2

No deduction of formwork is made where the beam meets the face of the pier.
(See SMM E20.16–M11.)

The deduction for the columns is a want, i.e. occurring at the boundary and therefore must be deducted. (See SMM General Rules 3.4.)

$$\underline{Girth\ of\ slab\ edge}$$

$$\begin{array}{r} 18·894 \\ 2/\ 9·644 \\ \hline 28·538 \\ \hline 57·076 \end{array}$$

57.08

Fmwk. for in situ conc edges of suspd. slabs, plain vert. hgt. ≤ 250 mm

SMM E20.3.1.2.0

The previous *to take* note for reinforcement will cover all of the reinforcement required.

To Take

1. Adj. for staircase well
2. Soil to finished ground level to measure with external works
3. Brickwork, dpc. etc above recess & top of slab
See note on p. 17.

These *to take* notes are out of order, but have been noted here due to lack of room on C.B. 17. (See cross references here and C.B. 17.)

Conc Basement 16

Asphalt

57·90 3·23	2ct. Mastic asp. tanking a.b. width > 300mm vert. 20mm thick on conc. fin. with a wood float, sub. covered.

SMM J20.1.4.1.1

Asphalt is measured the area in contact with the base. (See SMM J20.1–M3.)

Angle fillet

57·90	Mastic asp. a.b. int. ∠ fillet triangular 50×50mm

SMM J20.12.1.0.0

Recess at top of wall

4/2/½/ ext. gi. 57·900
−passings ½/0·103= 0·412
                    57·488

57·49 0·10	2ct. Mastic asp. tanking a.b. width ≤ 150mm. horiz. 20mm thick on conc sub covered.

SMM J20.1.1.1.1

4/2/½/ ℄ gi. 57·488
−passings ½/0·103= 0·412
                    57·076

57·08 0·15	Ditto width ≤ 150mm. vert. 20mm thick ditto.

SMM J20.1.1.1.1

Name of Project

Name of date

Name 17

Conc.   Basement   17

₤ girth of brick wall
ext. girth of conc. wall 57·900
+passing asp. 2/4/2/½/0·020 = 0·160
+   ,,   bwk 4/2/½/0·103 = 0·412
                              58·472

| 58·47 | |
| 3·23 | |

Bk. walls ½ B thick in
Class B Engineering bks.
in stretcher bond in
sulphate resistant c.m.
(1:3), vert. bdg. against
other work.

SMM F10.1.1.1.1

No working space is needed for
the brickwork as working space
has been measured for the full
depth of excavation for
formwork to the toe foundation.

Back filling to perimeter
girth of earthwk. supp⁺ 60·300
−passings 4/2/½/0·177 = 0·708
          ₤ girth 59·592
                   Depth
Av. exc depth 3·370
−blinding layer 0·050
−conc bed   0·100·0·150
                   3·220

59·59	
0·18	
3·22	

Filling to excavtns. av. thick
≯ 0·25m arising from excavtn.

SMM D20.9.2.1.0

+

Ddt disp. of excavtd. mat.
off site

SMM D20.8.3.1.0

NB. To take notes for top soil
and bwk added back to P. 15

6

Structural walls

Introduction

The measurement of structural walls may be carried out by a variety of methods, depending on the format of the bills of quantities to be adopted, e.g. elemental, trade, etc., although the approach for each is similar. However, before measurement can commence, plans of all floors, roof levels, elevations, sections and all necessary details will be required.

Approach

Measurement should be divided into manageable portions for each individual building or parts of a building and then further divided into:

1. external walls
2. internal walls
3. chimney breasts
4. chimney stacks

The measurement of fire surrounds, flue liners, chimney pots and cappings would normally be measured as a separate section and thus could be measured by a different taker-off.

Working from the whole to the part, measure walls over all openings and projections and then adjust for blank openings (not window or door openings which will be adjusted in that particular section), recesses, different types of facing bricks, projections and similar features.

All types of facing applied to the external walls, e.g. rendering, tile hanging, etc., are measured in this section.

Generally

Brickwork and blockwork are generally measured as complete items on the centre line of the wall with the nominal thickness stated, see SMM F10–M1 and D1. Every description must include the supplementary information requirements (see SMM F10–S1–S5) as necessary, in addition to the usual classification terms. Depending on the complexities of the project, these requirements may be more conveniently covered by a heading or preamble clauses in the bills of quantities.

S1 – Kind, quality and size of bricks, etc., usually covered by stating a manufacturer's name and size of bricks.

S2 – Type of bond, e.g. stretcher, English or Flemish bond; this may affect the number of facing bricks used.

S3 – Composition of bricks and mortar, e.g. cement mortar 1:3 (one part cement and three parts fine aggregate) or gauged mortar 1:1:6 (one part cement, one part lime and six parts fine aggregate).

S4 – Type of pointing; this is the treatment of the face of the mortar joints. There are different finishes, e.g. weather struck joint, flush joint, bucket handle joint, etc., but, irespective of the type of finish, they are all executed by one of the two following methods:

1. 'As work proceeds'. The bricklayer will complete the face of the mortar joints to the required finish at some convenient time before the mortar has set. This is a relatively cheap process. The code of practice relating to brickwork describes this process as 'jointing', but as SMM7 F10–S4 refers to 'pointing' only, it is described in the following examples as 'pointing with a ... joint as work proceeds'.

2. 'As a separate operation'. The bricklayer will rake out the mortar joints to a depth of approximately 12 mm while the mortar is still green and then, after the brickwork is finished, but before the scaffolding is dismantled, the joints will be filled with mortar and the face treated. This gives the brickwork a uniform appearance but is a very much more expensive method of pointing.

The term 'facework' all includes any work in bricks and block finished fair, see SMM F10–D2. Thus, facework not only includes work built in facing bricks and pointed but also walls built in common bricks, fair faced and pointed. The term 'fair faced' means providing a good finished appearance to the face of common brickwork. The bricklayer will select the common bricks, choose the best face of the brick, and take extra care in laying and pointing.

Measurement

Measure the walls up to a general level, e.g. roof plate level, overall openings, recesses, etc., and add for gable walls, all adjustments being made later. The taker-off must decide whether or not to measure the walls over openings. Generally the decision depends on whether the wall is to be supported on a lintel over the opening (with adjustments being made with the doors or windows sections), or if the doors have storey height frames, i.e. floor to ceiling, in which case there is no need to measure the wall over only to be deducted in full at a later stage. These openings are termed nett openings and they should be noted on the drawing and the taking-off. Deductions for lintels, etc., are made as regards height to the extent only of full brick or block courses displaced and as regards depth to the extent only of full half brick beds displaced, see SMM F10–M3.

For example, with regard to Figure 6.1, blockwork is not deducted as the lintel does not displace a full course. If the lintel was 225 mm high a full block would be displaced and an appropriate deduction would be made.

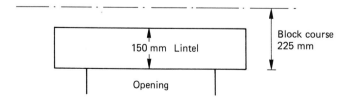

Figure 6.1   *Elevation*

When measuring chimney breasts and chimney stacks, the brickwork is measured as if it is solid unless the flue exceeds $0.25\,m^2$, see SMM F10–M2, and Figure 6.2.

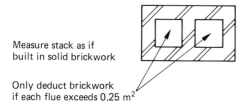

Measure stack as if
built in solid brickwork

Only deduct brickwork
if each flue exceeds $0.25\,m^2$

Figure 6.2 *Plan of chimney stack*

Isolated piers are measured as such only if dimensions are as shown in Figure 6.3, see SMM F10.2–D8, otherwise they are measured as a wall, see Figure 6.4.

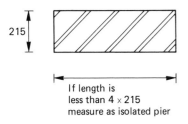

215

If length is
less than 4 × 215
measure as isolated pier

Figure 6.3 *Plan*

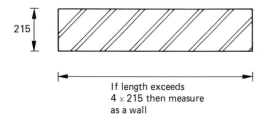

215

If length exceeds
4 × 215 then measure
as a wall

Figure 6.4 *Plan*

Attached piers are measured as projections if dimensions are as Figure 6.5, see SMM F10.5–D9; if not, measure as a wall of combined thickness of attached pier and wall, as Figure 6.6.

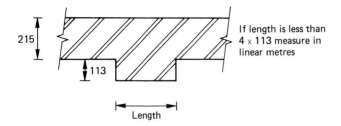

Figure 6.5   *Plan*

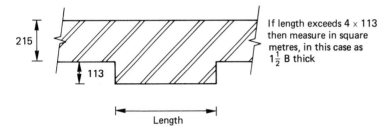

Figure 6.6   *Plan*

Flow chart for structural walls

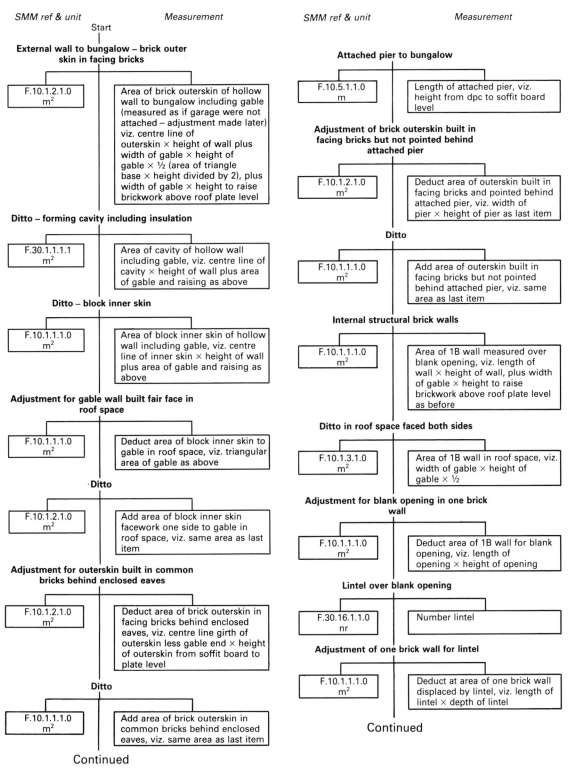

SMM ref & unit   *Measurement*

Start

External wall to bungalow – brick outer skin in facing bricks

F.10.1.2.1.0
m²

Area of brick outerskin of hollow wall to bungalow including gable (measured as if garage were not attached – adjustment made later) viz. centre line of outerskin × height of wall plus width of gable × height of gable × ½ (area of triangle base × height divided by 2), plus width of gable × height to raise brickwork above roof plate level

Ditto – forming cavity including insulation

F.30.1.1.1.1
m²

Area of cavity of hollow wall including gable, viz. centre line of cavity × height of wall plus area of gable and raising as above

Ditto – block inner skin

F.10.1.1.1.0
m²

Area of block inner skin of hollow wall including gable, viz. centre line of inner skin × height of wall plus area of gable and raising as above

Adjustment for gable wall built fair face in roof space

F.10.1.1.1.0
m²

Deduct area of block inner skin to gable in roof space, viz. triangular area of gable as above

·Ditto

F.10.1.2.1.0
m²

Add area of block inner skin facework one side to gable in roof space, viz. same area as last item

Adjustment for outerskin built in common bricks behind enclosed eaves

F.10.1.2.1.0
m²

Deduct area of brick outerskin in facing bricks behind enclosed eaves, viz. centre line girth of outerskin less gable end × height of outerskin from soffit board to plate level

Ditto

F.10.1.1.1.0
m²

Add area of brick outerskin in common bricks behind enclosed eaves, viz. same area as last item

Continued

SMM ref & unit   *Measurement*

Attached pier to bungalow

F.10.5.1.1.0
m

Length of attached pier, viz. height from dpc to soffit board level

Adjustment of brick outerskin built in facing bricks but not pointed behind attached pier

F.10.1.2.1.0
m²

Deduct area of outerskin built in facing bricks and pointed behind attached pier, viz. width of pier × height of pier as last item

Ditto

F.10.1.1.1.0
m²

Add area of outerskin built in facing bricks but not pointed behind attached pier, viz. same area as last item

Internal structural brick walls

F.10.1.1.1.0
m²

Area of 1B wall measured over blank opening, viz. length of wall × height of wall, plus width of gable × height to raise brickwork above roof plate level as before

Ditto in roof space faced both sides

F.10.1.3.1.0
m²

Area of 1B wall in roof space, viz. width of gable × height of gable × ½

Adjustment for blank opening in one brick wall

F.10.1.1.1.0
m²

Deduct area of 1B wall for blank opening, viz. length of opening × height of opening

Lintel over blank opening

F.30.16.1.1.0
nr

Number lintel

Adjustment of one brick wall for lintel

F.10.1.1.1.0
m²

Deduct at area of one brick wall displaced by lintel, viz. length of lintel × depth of lintel

Continued

Flow chart for structural walls – continued

SMM ref & unit	Measurement

Continued
Chimney breast ground floor

F.10.1.1.1.0 m²	Area of chimney breast – described as brick wall 2½B thick, viz. length of breast × height of breast (ground floor)

Adjustment of one brick wall for chimney breast

F.10.1.1.1.0 m²	Deduct area of one brick wall displaced by chimney breast, viz. same area as last item

Chimney breast roof space

F.10.5.1.1.0 m	Length of chimney breast in roof space, viz. average height of chimney breast attached to one brick wall in roof space

Adjustment of one brick wall pointed both sides in roof space behind chimney breast

F.10.1.3.1.0 m²	Deduct area of one brick wall facework both sides, viz. width of chimney breast in roof space × average height of chimney breast

Ditto

F.10.1.2.1.0 m²	Add area of one brick wall facework one side, viz. same area as above item

Chimney stack

F.10.4.3.1.0 m²	Area of chimney stack above roof level, viz. width of stack × average height of chimney stack

Damp proof course in chimney stack

F.30.2.2.3.0 m²	Plan area of stack, viz. width of stack × depth of stack

Chimney capping flue linings etc

To take note measure with fires and vents section	Not measured in this example

Continued

SMM ref & unit	Measurement

External brick wall in facing bricks to garage

F.10.1.3.1.0 m²	Area of half brick wall to garage, viz. girth of garage wall × height of garage wall

Attached piers to garage

F.10.5.1.1.0 m	Length of attached piers, viz. same height as garage × 2

Adjustment of garage wall not pointed both sides behind attached pier

F.10.1.3.1.0 m²	Deduct area of garage wall facework both sides behind attached piers, viz. width of pier × height of pier as last item × 2

Ditto

F.10.1.2.1.0 m²	Add area of garage wall facework one side behind attached piers, viz. same area as last item

Brick screen wall

F.10.1.3.1.0 m²	Area of half brick screen wall, viz. length of screen wall × height of screen wall

Attached pier to screen wall

F.10.5.1.1.0 m	Length of attached pier, viz. same height as screen wall

Adjustment of screen wall not pointed both sides behind attached pier

F.10.1.3.1.0 m²	Deduct area of screen wall facework both sides behind attached pier, viz. width of pier × height of pier as last item

Ditto

F.10.1.2.1.0 m²	Add area of screen wall facework one side behind attached pier, viz. same area as last item

Continued

Flow chart for structural walls – continued

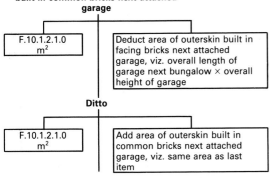

Continued

**Adjustment for outerskin of bungalow
built in common bricks next attached
garage**

F.10.1.2.1.0
m²

Deduct area of outerskin built in facing bricks next attached garage, viz. overall length of garage next bungalow × overall height of garage

Ditto

F.10.1.2.1.0
m²

Add area of outerskin built in common bricks next attached garage, viz. same area as last item

Check drawings and taking-off list for outstanding items

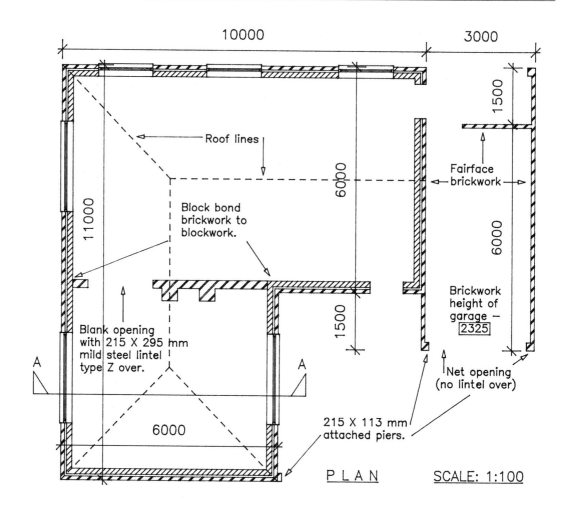

PLAN   SCALE: 1:100

STRUCTURAL WALLS

SPECIFICATION

1. External walls: 303 mm hollow wall comprising half brick outer skin in Messrs X Multi—colour facing bricks in stretcher bond in gauged mortar (1:1:6) pointing with a weathered joint as work proceeds. 50 mm cavity with 'Hemax' 50 stainless steel wall ties 225 mm long at 900 mm centres horizontally & 450 mm centres vertically, staggered.   (Contd.)

SPECIFICATION (Cont.)
25 mm 'Dri Therm' rigid insulation, and 150 mm
Thermalite Turbo Blocks in gauged mortar (1:1:6).
2. Roof space: All exposed brickwork and blockwork to
   be built fairfaced and pointing with a flush joint
   as work proceeds.

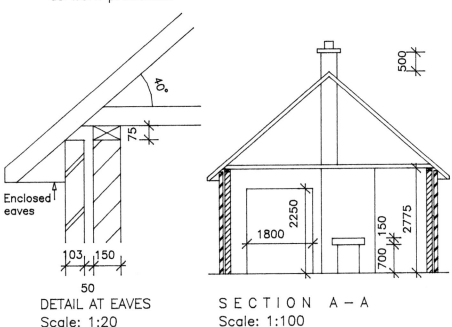

DETAIL AT EAVES
Scale: 1:20

SECTION A – A
Scale: 1:100

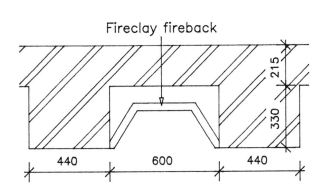

SECTIONAL PLAN OF CHIMNEY BREAST
Scale: 1:20

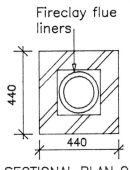

SECTIONAL PLAN OF
CHIMNEY STACK
ABOVE ROOF LEVEL
Scale: 1:20

Drawing Number

Name of Project

date

Name 1

Structural Walls 1

Taking Off List

1. External walls

2. Adjustment of fair faced work in roof space.

3. Adjustment of faced work at eaves

4. Attached piers

5. Internal walls including walls in roof space.

6. Blank opening adjustment

7. Chimney breast

8. Chimney stack

9. Garage

10. Screen wall

11. Adjustment of external wall next garage

Page Nrs.

1	11
2	12
3	13
4	14
5	15
6	16
7	17
8	18
9	19
10	20

Name 2   date   Name   of   Project.

Struct. Walls 2

<u>Ext. walls</u>

The approach for this example is to measure the bungalow ignoring the attached garage and screen wall completely. The garage etc. is then measured later and adjustments made to the dividing wall.

₵ girth outerskin

	10·000
	11·000
2/	21·000

ext. girth. = 42·000

ext. corners = 5
− int. ,, = 1
passing timesing. 4
−passings ⁴/₂/½/0·103 0·412
₵ girth %skin :41· 588

To calculate the external girth of brickwork of an 'L' shaped building, add together the two longest overall dimensions (in this case ignoring the garage) which are at right angles to each other and times by two.

<u>Height.</u>
dpc to soffit. 2·775
− wall plate 0·075
2·700

The superstructure is measured from the d.p.c. level.

<u>Gable Wall</u>

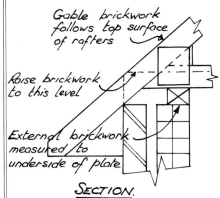

Gable brickwork
follows top surface
of rafters

Raise brickwork
to this level

External brickwork
measured to
underside of plate

<u>SECTION.</u>

The height of the external wall generally has been measured to the underside of the wall plate. Therefore the measurement of the gable wall will comprise the raising of

Struct. Walls 3

Ext Walls (Ctd)

Gable wall (Ctd)

the general wall height by 150 mm, and then a triangle following the shape of the top line of the rafters. (See drawing on Struct. Walls 2.)

$$\frac{Hgt. \ of \ gable}{}$$
= Tan 40° × half span
= Tan 40° × 3·000
= 2·517

The height of the triangle is calculated by trigonometry. Pitch of roof is 40°.

The outerskin, cavity and inner-skin will all be the same size.

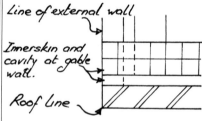

Line of external wall

Innerskin and cavity at gable wall.

Roof Line

PLAN OF GABLE WALL

41·59
2·70
6·00
½/ 0·15
6·00
2·52

Walls facework one side h.b. thick, vert. Messrs 'X' multi. col. fcg. bks. (gable end in stretcher bond (raising in g.m (1:1:6) & (gable end ptg. wi. a (triangle weathered struck jt. a.w.p.

SMM F10.1.2.1.0

The term *walls* includes skins of hollow wall. (See SMM F10–D4.)

₤ girth cavity
₤ girth o/skin. a.b.   41·588
– passings a.b.   0·412
– cavity 4/2/½/ 0·050   0·200·0·612
40·976

Name of Project

date

Name 3

Struct. Walls 4.

<u>Ext. walls (Ctd)</u>
<u>Cavity (Ctd)</u>

	40.98	Forming cavity in h.w.
	2.70	50 mm wide inc. "Hemax
	6.00	50" stainless (gable end
	0.15	steel wall ties (raising
½/	6.00	225 mm long (gable end
	2.52	spaced at 900mm triangle

SMM F30.1.1.1.1

The requirements of clauses
SMM F30.1–S2 & S3 have to be
complied with by including
the supplementary information
either in the description or
more probably in the
Preambles.

centres horiz. & 450 mm
centres vert. staggered,
rigid "Dri Therm" insul.
25 mm thick fixed with
plastic clips to wall
ties.

<u>& girth innerskin.</u>
& girth cavity ab: 40.976
–cavity passing ⌀b.: 0.200
–innerskin 4/2/½/0.150 : 0.600 0.800

40.176

	40.18	Walls 150 mm thick, vert.
	2.70	in Thermalite Turbo
	6.00	blks. in stretcher (gable end
	0.15	bond in g. m. (raising
2/	6.00	(1:1:6) (gable end
	2.52	(triangle

SMM F10.1.1.1.0

Name of Project.

date

Name 4

Struct. Walls 5

Adj. for facework in roof space
                    Gable wall.

The exposed face of walls in
the roof space is to be
finished fair.

±/	6·00	Ddt walls 150mm thick
	2·52	in Thermalite blks a.b.

SMM F10.1.1.1.0

&

Add walls facework one
side 150mm thick in
'Thermalite Turbo' blks.
in stretcher bond in
g.m. (1:1:6) & ptg. with
a flush jt. a.w.p.

SMM F10.1.2.1.0

Adj. of outerskin behind
enclosed eaves.

Ext. girth a.b. 42·000
−gable wall   6·000
                    36·000
−passings 2/2/½/ 0·103 0·206
              ₤ girth. 35·794

Outerskin in
common bricks

225 mm

Outerskin in facing
bricks

SECTION.

Struct. Walls 6

Adj. of outerskin behind eaves (Ctd)

35·79	<u>Ddt</u> walls facewk. one side h. b. thick in Messrs 'X' fcg. bks. a.b.	SMM F10.1.2.1.0
0·23		

&

35·79		

<u>Add</u> walls h. b. thick. vert. in c.b. in stretcher bond in g.m. (1:1:6)

SMM F10.1.1.1.0

Attached pier

NB. Pier built to soffit board level.

	Hgt.
Outerskin	2·700
- soff. depth	0·225
	2·475

2·48	Projections 215 mm. wide & 113 mm. proj. vert. in Messrs 'X' multi - col. fcg. bks. in English bond in g.m. (1:1:6) & ptg. with a weathered struck jt. a.w.p.	SMM F10.5.1.1.0

Labours to returns are deemed to be included. (See SMM F10–C1(f).)

Name of Project

Name 6 date

Struct. Walls 7.

Attached pier (Ctd)

0·22	*Ddt* walls facework one
2·48	side in Messrs 'X' fcg
	bks a.b.

SMM F10.1.2.1.0

The wall behind the attached
pier is built in facing bricks
but not pointed on face.
Probably would not be measured
in practice.

&

Add walls h.b. thick, vert
in Messrs 'X' fcg. bks.
in stretcher bond in
in g.m. (1:1:6)

SMM F10.1.1.1.0

Internal structural wall.

ext. dims.   6·000
- ext walls ²/0·303·0·606
                    5·394

5·39	Walls one bk.	(underside
2·70	thick vert. in	(plate
6·00	c.b. in English	(raising
0·15	bond in g.m.	(above
	(1:1:6)	(plate a.b

SMM F10.1.1.1.0

Name 7   date   Name of Project

Struct. Walls 8.

Int. struct. wall (Ctd)
wall in rf. space.

½/6.00
2.52

Walls facework both sides
one bk. thick in c.b.
in English bond in
g.m. (1:1:6) & ptg both
sides with a flush jt.
a.w.p.

SMM F10.1.3.1.0

There does not appear to be a
classification in SMM for the
measurement of block bonding
at the junction of the brick
internal wall and block
external wall. The
classification requirements in
SMM.F10.1.0.0.2 refer to
superficial items and F10.25
refers to existing work.

Adj. of blank opg.

Sometimes an opening is
required through a wall which
will not have a door or window
set in it. This is called a
blank opening.
The adjustment of the
structure will not be carried
out in the usual way, i.e.
structure etc. deducted for the
opening when the window or door
is measured.
Therefore in the case of a
blank opening each taker-off
is responsible for adjusting
his own dimensions.
These openings are usually
indicated by having 'blank' or
'net' opening written over the
opening on the drawings.

Name 8   date   Name of   Name of Project.

Struct. Walls 9

Adj. of blank opg. (Ctd)

1·80	Ddt 1B wall in c.b.	SMM F10.1.1.1.0
2·25	a.b.	
2·25	(lintol	
0·25		

Lintol

opg.   1·800
ends 2/0·225·0·450
          2·250

1	Proprietary gal. m.s lintol 215 mm wide × 215 mm hi. & 2250 mm lg. Type 'Z' manufactured by Messrs 'Y' & b.i. to bwk.	SMM F30.16.1.1.0

bwk. adj.
added back.

Chimney breast

The chimney breast is measured gross, i.e. over the fireplace opening. The adjustments for the fireplace opening and measurement of flue linings etc. will be made in the Fires and Vents section of the taking-off.

Struct. Walls 10.

Chimney breast (Ctd)

The brickwork in chimney breasts and chimney stacks is measured solid as if there are no flues in them. Brickwork will only be deducted if the flue area exceeds 0.25 m². (See SMM F10–M2(b).)

The usual cross-sectional area of a domestic flue is $0.215 \times 0.215 = 0.05$ m².

Height

Grd. flr.   2·775
into roof space : 0·075
2·850

Length

0·440
0·600
0·440
1·480

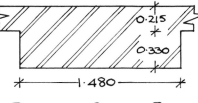

PLAN OF CHIMNEY BREAST.

4×0.330 (proj.) = 1·320
As length 1·480 > four
times proj. 1·320 then
measure chimney breast
as o wall.

Thickness

Wall	1B
proj.	1½B
	2½B

SMM F10.5–D9 defines a projection. If the length of the projection (attached pier) ≤ four times their thickness then the work is measured in linear metres as a projection. If the length exceeds four times the thickness then it is measured in square metres as a wall of that thickness plus the thickness of the backing wall.

1·48	
2·85	

Walls 2½ B thick vert. in c.b.
in Eng. bond in g.m.(1:1:6)

&

Dd‡ Walls 1B thick in c.b.a.b

SMM F10.1.1.1.0

SMM F10.1.1.1.0

<u>Struct. Walls **11**</u>

<u>Chimney breast (Ctd)</u>
<u>Roof space</u>
Hgt.
ext. walls 2·700
rf. plate etc. 0·150
to apex of △ in rf. 2·517
5·367
- breast grd. flr. 2·850
2·517

Adj. of △ at top of chimney
breast.

$$X = \tan 40° \times \frac{0·440}{2} = 0·185$$

$$- \quad av = \tfrac{1}{2}/0·185 = 0·093$$

2·424

<u>Width</u>
0·440

As width 0·440 ∠ four times
proj. 1·320 then measure
chimney breast as a projection.

2·42	Proj. 440mm wide ⋋ 330mm deep vert. in c.b. in Eng. bond in g.m (1:1:6) built fair face ⋋ ptg. with a flush joint a.w.p.

SMM F10.5.1.1.0

0·500 Chimney Stack

2·517

0·440
Chimney Breast.

ELEVATION.

0·330

0·440

Struct. Walls 12

Chimney Breast (Ctd)
Adj. of wall behind chimney breast.

0·44	Ddt walls facework both sides 1B thick in c.b. a.b.	SMM F10.1.3.1.0
2·42		

&

	Add walls facework one side 1B thick. vert. in c.b. in Eng. bond in g.m. (1:1:8) & ptg. with a flush jt. a.w.p.	SMM F10.1.2.1.0

Chimney Stack

	Hgt
Above roof	0·500
+ Adj. of △ at top of rf.	0·093
	0·593

0·44	Chimney stacks facework both sides 2B thick vert. in Messrs X fcg. bks. a.b. in Eng. bond in c.m. (1:3) with joints raked out 10mm dp & ptd. in g.m.(1:1:6) & finished with a weather struck jt. as a separate operation.	SMM F10.4.3.1.0 The chimney stack is usually built in cement mortar due to its exposed position but the mortar will appear to give the faced brickwork a different colour from that built in gauged mortar. Therefore the joints of brickwork are raked out and pointed in gauged mortar to give the chimney stack a similar appearance.
0·59		

Name of Project

date

Name 12

Struct. Walls 13.

Chimney Stack (Ctd)

There is no clause in SMM to measure pointing ends of the chimney stack as the labour is deemed to be included. (See SMM F10.C1(f).)

0·44	Dp.c width > 225 mm. horiz bit. felt as spec. single layer bedded in c.m. (1:3)
0·44	

SMM F30.2.2.3.0

No deduction is made for the flue. (See SMM F30–M3.)

To Take
Fire surround, flue linings, capping etc in Fires & Vents.

Garage
$\mathcal{L}$ girth of wall

back wall	3·000
side ..	6·000
	9·000
-passing ½/½/0·103 =	0·103
	8·897
front flank wall	1·500
$\mathcal{L}$ girth	10·397

10·40	Walls facework both sides h.b. thick vert. in Messrs X fcg. bks. a.b. in stretcher bond in g.m. (1:1:6) & ptg one side with a weather struck joint & other side with a flush jt. a.w.p.
2·33	

SMM F10.1.3.1.0

<u>Struct. Walls 14.</u>

<u>Garage (Ctd)</u>
<u>Attached piers.</u>

| 2/ | 2.33 |
| | 2.33 |

Proj. 215 mm wide & 113 mm
deep, vert. in   (Screen wall
Messrs 'X' fcg bks
a.b. in Eng. bond in g.m.
(1:1:6) & ptg. with a
weather struck jt. a.w.p.

SMM F10.5.1.1.0

<u>Adj.</u> of walls behind piers

| 2/ | 0.22 |
| | 2.33 |

<u>Ddt</u>   walls facework both
sides in Messrs 'X' fcg bks.
a.b. ptg. one side with a
weather struck jt. & other
side with a flush jt. a.w.p.

SMM F10.1.3.1.0

&

<u>Add</u> walls facework one
side in Messrs 'X' fcg bks
a.b. ptg. one side with
a weather struck jt. a.w.p.

SMM F10.1.2.1.0

<u>Struct. Walls 15.</u>

<u>Screen walls.</u>

1·50	Walls facework both sides	SMM F10.1.3.1.0
2·33	h.b. thick, vert. in Messrs 'X' fcg. bks a.b. in stretcher bond in g.m (1:1:6) & ptg. both sides with a weather struck jt. a.w.p.	

N.B. no coping to top
of wall.
pier added back
to struct. walls 14.

<u>Adj. of wall behind pier</u>

0·22	<u>Ddt</u>. ditto	SMM F10.1.3.1.0
2·33		

&

<u>Add</u> walls facework one side h.b. thick, vert. in Messrs 'X' fcg. bks. a.b. in stretcher bond in g.m. (1:1:6) & ptg with a weather struct jt a.w.p.	SMM F10.1.2.1.0

Struct. Walls 16

<u>Adj. of ext. wall of</u>
<u>Bungalow next garage.</u>

<u>Length</u>
6·000
− 1·500
4·500

<u>Hgt.</u>
garage bwk. 2·325
+ roof   0·225
2·550

4·50	
2·55	

<u>Dolt</u> walls facework (outerskin
one side h.b. thick, vert.
in Messrs 'X' fcg. bks. a.b.
& ptg. with a weather
struct jt. a.w.p.

SMM F10.1.2.1.0

&

<u>Add ditto</u> in c.b. in stretcher
bond in g.m. (1:1:6) & ptg.
with a flush jt. a.w.p.

SMM F10.1.2.1.0

Name of Project.

date

Name 16

Name

Non-structural walls

Generally

The notes included in Chapter 6 apply equally to this section.

If blockwork partitions are to have a faced finish then cutting of the blocks may be required, in which case SMM F10–S5 clause must be complied with. Special blocks may be used at reveals, angles, intersections, etc., to set out the required bond, in which case they must be measured in linear metres, see SMM F10.11.

The taker-off should also measure any restraints at ends, e.g. ties, struts, etc., and at bottom and heads of partitions, e.g. plates and noggins, and any damp proof courses which may not be shown on the foundation plan.

Measurement

Where appropriate, the use of a schedule is advocated to collate dimensions or at least side casts for partitions of the same type. Each partition should be given a unique reference (see example) and then marked on the drawing as measured.

It should be noted that the height of the partitions is measured from the structural floor level and not finished floor level.

Flow chart for non-structural walls

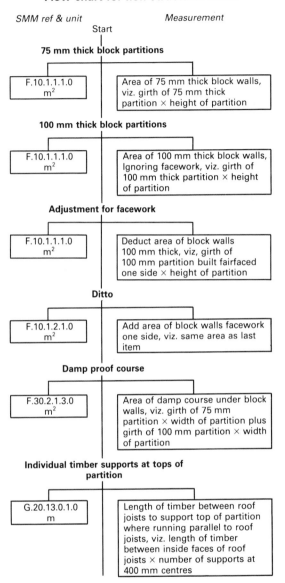

SMM ref & unit   *Measurement*

Start

75 mm thick block partitions

| F.10.1.1.1.0
m² | Area of 75 mm thick block walls,
viz. girth of 75 mm thick
partition × height of partition |

100 mm thick block partitions

| F.10.1.1.1.0
m² | Area of 100 mm thick block walls,
Ignoring facework, viz. girth of
100 mm thick partition × height
of partition |

Adjustment for facework

| F.10.1.1.1.0
m² | Deduct area of block walls
100 mm thick, viz, girth of
100 mm partition built fairfaced
one side × height of partition |

Ditto

| F.10.1.2.1.0
m² | Add area of block walls facework
one side, viz. same area as last
item |

Damp proof course

| F.30.2.1.3.0
m² | Area of damp course under block
walls, viz. girth of 75 mm
partition × width of partition plus
girth of 100 mm partition × width
of partition |

**Individual timber supports at tops of
partition**

| G.20.13.0.1.0
m | Length of timber between roof
joists to support top of partition
where running parallel to roof
joists, viz. length of timber
between inside faces of roof
joists × number of supports at
400 mm centres |

Check drawings and taking-off list for outstanding items

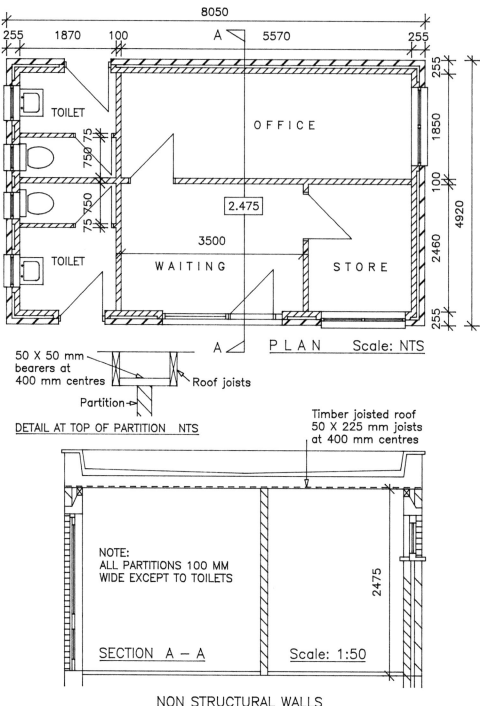

8050

255 1870 100 A 5570 255

255

TOILET

1850

OFFICE

75 75

750 75

100

2.475

4920

75 750

3500

2460

TOILET

WAITING

STORE

255

A   PLAN   Scale: NTS

50 X 50 mm
bearers at
400 mm centres

Roof joists

Partition

DETAIL AT TOP OF PARTITION   NTS

Timber joisted roof
50 X 225 mm joists
at 400 mm centres

NOTE:
ALL PARTITIONS 100 MM
WIDE EXCEPT TO TOILETS

2475

SECTION   A – A

Scale: 1:50

NON STRUCTURAL WALLS

<u>Non Structural Walls</u> **1**

<u>Taking Off List</u>.

1. Block partitions

2. Adjustment for fair face blockwork

3. D.p.c

4. Restraints at tops of partitions

Drawing Number

Name of Project

Name / date

<u>Page Nrs</u>.
~~1~~
~~2~~
~~3~~
~~4~~
~~5~~
6
7
8
9
10

Name 2   Name of Project.   date

Non Struct. Walls 2

Partitions

Rm.1		3
2		
7		
6	5	4

PLAN.

¢ girths
75 mm. part.
part 1/2 : 1·870
+   „   6/7 : 1·870/
                3·740

100 mm. part.
Horiz.- part. 2/7 : 1·870
        part : 0·100
   „   3/4&5: 5·570
                7·540
Vert. part 1&2/3 1·850
   „   4/5 2·460
   „   5/6&7 2·460
            14·310

Check
o/a dims   8·050
– walls 2/0·255: 0·510
                7·540

It is helpful to draw a line diagram on the dimensions and number the rooms. Each partition then has a unique reference, e.g. 1/2 or 1 & 2/3. In large projects the measurement may involve many different kinds of partitions (e.g. both fixed and demountable), different heights etc. and it is necessary to identify what has been measured. It helps to mark the architect's drawing as the partitions are measured.

The measurement convention used is to collate together on waste the centre line girth of all like partitions of a like height by taking the horizontal partitions, i.e. going from left to right on the drawing first, and then the vertical partitions, i.e. going top to bottom of the drawing, although sometimes it is easier to measure room by room.

It is useful to check major dimensions which have been built up piecemeal before commencing taking-off.

Non Struct. Walls 3

Parts (Ctd)

3·74		Walls 75mm thick vert. in Messrs 'X' lightweight blks. in stretcher bond in g.m (1:1:6)	SMM F10.1.1.1.0
2·48			

14·31		Walls 100mm ditto.	SMM F10.1.1.1.0
2·48			

Adj. for facework Rm 4

	Rm 3 5·570
Rm. 5 – 3·500	
part. 0·100	3·600
part. 3/4	1·970
,, 4/5	2·460
	4·430

4·43		Ddt walls 100mm ditto	SMM F10.1.1.1.0
2·48			

&

		Add walls facework one side 100mm thick vert. in Messrs 'X' lightweight blks. in str. bond in g.m.(1:1:6) & ptg. wi. a flush jt. a.w.p.	SMM F10.1.2.1.0

The measurement of facework on block walls or partitions is a grey area which falls between

Non Struct. Walls 4

<u>Adj. for f.wk. (Ctd)</u>

<u>To take</u>
Check if facework in
room 4 on external walls
is measured in the
structural wall section.

<u>D.p.c</u>

3·74	
0·08	
14·31	
0·10	

Pitch Polymer (75mm part
d.p.c. width ≤ 225
mm. horiz. single (100mm ,,
layer bedded in
g.m. (1:1:6). 150 mm.
laps (meas. nett)

<u>Top restraints</u>

centres of rf. jsts 0·400
- jsts 7½/0·050·   0·050
noggin length   0·350

<u>Nr. of noggins</u>
part 1&2/3·0450 | 1·850
4 + 1 end
c/f 5 Nr.

two taking off sections, i.e.
non-structural walls and
internal finishings, and it
must be clearly defined who is
to measure it. Usually as
facework is included in the
description of the blockwork
it is the responsibility of
the partitions taker-off to
measure the facework and so he
must consult the internal
finishings schedule to find
out which rooms are finished
fair face.

SMM F30.2.1.3.0

The non-structural ground
floor partitions may not be
shown on the sub-structure
drawing and therefore it is
necessary to liaise with the
sub-structure taker-off.

Where there is a timber upper
floor or timber roof it may be
necessary to have bearers in
between the joists to act as a
form of restraint to the top
of the partition. This occurs
where the partitions run
parallel with the joists and
fall in between two joists.

Name of Project.

Name of   date

Name 4

Name 5   Name   date   Name of Project.

<u>Non - Struct. Walls 5</u>

<u>*Top restraints (Ctd)*</u>

Where the partitions run at right angles to the joists they are restrained by the joists.

<u>*Nr. of noggins*</u> b/f 5 Nr

part. 4/5·0·450) 2·460

5+1 rem+1 end 7 ·

part. 5/6+7 as part. 4/5: 7 ·

19 ·

19/0·35

50 × 50 mm Swn. Tan. swd.
individual support.

SMM G20.13.0.1.0

Stud partitions

Generally

Each component of stud partitions is measured in linear metres and described as 'wall or partition members' giving cross-section dimensions of the components, see SMM G20.7.*.1.

The measurement of each component must include sufficient allowance of timber to enable the carpenter to make any joints the designer deems necessary, although the work is deemed to include labours on items of timber, see SMM G20–C1. The method of jointing and method of fixing must be stated where not at the discretion of the contractor, see SMM G20–S2 and S9 respectively, see Figure 8.1.

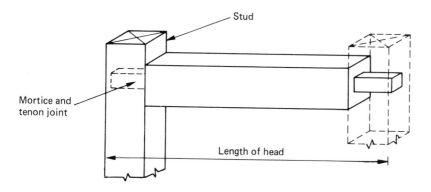

Figure 8.1 *Sketch to show mortice and tenon joint*

Adjustments for door openings are made at the time the partition is measured and consideration should be given to the measurement of any necessary restraints.

Measurement

Notes made for non-structural walls (Chapter 7) apply equally to this section.

Flow chart for stud partitions

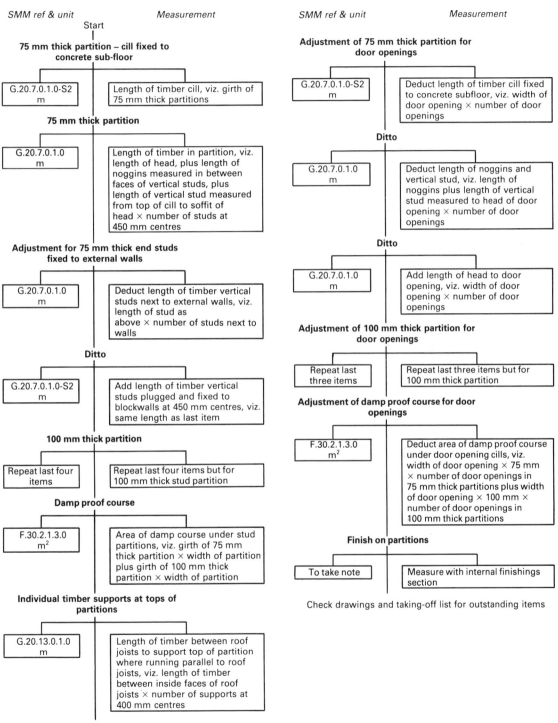

Continued

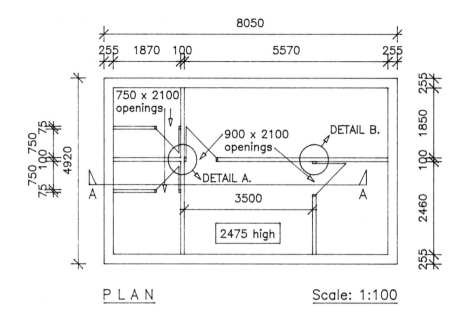

P L A N   Scale: 1:100

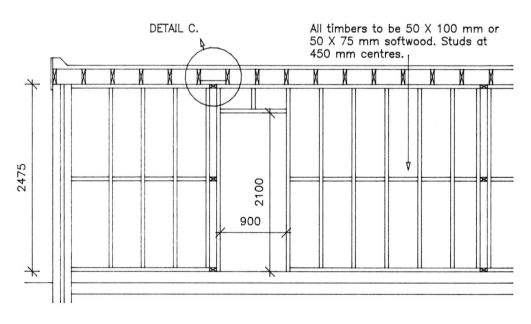

P A R T   S E C T I O N   A — A   Scale: 1:50

N O N   S T R U C T U R A L   S T U D   P A R T I T I O N S

DETAILS SHOWING ARRANGEMENT OF STUDS AT INTERSECTIONS

Scale: 1:10

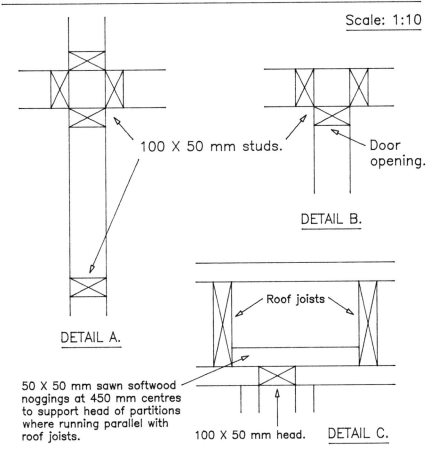

100 X 50 mm studs.

Door opening.

DETAIL B.

DETAIL A.

Roof joists

50 X 50 mm sawn softwood noggings at 450 mm centres to support head of partitions where running parallel with roof joists.

100 X 50 mm head.

DETAIL C.

SPECIFICATION
1. Softwood to be prime grade timber pressure impregnated with Tanalith C.
2. Cills to have Pitch Polymer damp proof course under and plugged and fixed at 450 mm centres to concrete floor.
3. Vertical studs next to solid walls to be plugged and fixed at 450 mm centres.

NON STRUCTURAL STUD PARTITIONING

Stud Partitions 1

Taking Off List

1. Stud partitions
   a. Cill
   b. Head
   c. Studs
   d. Noggins

2. Adjustment of fixing studs.

3. D.p.c.

4. Restraints at tops of partitions.

5. Adjustment of openings.

Drawing Number

Name of Project

date

Name 1

Page Nrs
1
2
3
4
5
6
7
8
9
10

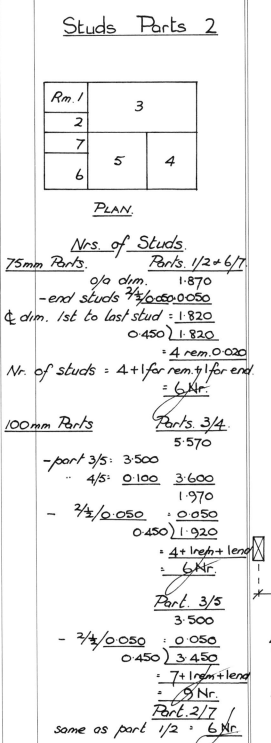

A unique reference can be provided for each partition by lettering or numbering the rooms on the drawings or on a sketch in the dimensions, e.g. partition between rooms 1 & 3 and 2 & 3 can be referenced 1 & 2/3.

As each partition is measured, it should be coloured in on the drawing.

To work out the number of studs from the information shown on the drawings it is necessary to work out the horizontal dimension between the centres of the first and last studs. This dimension is then divided by the centres of studs. An extra stud is added for any remainder (see later note) and then another stud is added to give the total number of vertical studs required.

Stud Parts. 3.

Nrs. of Studs(Ctd)
100mm thick parts.(Ctd)
Part 1/3
1·850
− Rm 2 : 0·750
part. : 0·075·0·825
1·025
− 2/½ 0·050 : 0·050
0·450)0·975
: 2 + 1 rem + 1 end
: 4 Nr.

Part. 2/3
0·750
− 2/½ 0·050 : 0·050
0·450)0·700
1 + 1 rem + 1 end
: 3 Nr.

Part 4/5
2·460
− 2/½ 0·050 : 0·050
0·450)2·410
5 + 1 rem + 1 end
: 7 Nr.

Part. 5/6
2·460
− Rm. 7 & part. a.b : 0·825
1·635
− 2/½ 0·050 : 0·050
0·450)1·585
3 + 1 rem + 1 end
: 5 Nr.

Part 5/7
same as part 2/3 : 3 Nr.

The centres of the vertical studs are dictated by the thickness of the finishings applied to create the partition. The specified centres of studs are the maximum distance the finishings will span. Therefore an additional stud is added for any remainder to ensure that the specified centres are not exceeded.

When calculating the number of vertical studs the length of partitions used must be that of the smallest room, thus ensuring that the studs are not undermeasured, e.g. use widths of Rooms 1 and 2 separately and not width of adjoining Room 3 when calculating studs for partition 1 & 2/3.

Name of Project.

date

Name 3

Stud Parts. 4.

Cills.

$^2/_{1.87}$	75 × 50 mm Sawn	(part 1/2
	Tan. swd. wall or	(·· 6/7
	part. members	
	p.& fxd. to conc.	
	at 450 mm centres	

SMM G20.7.0.1.0 – S2.

If the architect requires a specific method of fixing then this must be described, otherwise the fixing will be left to the contractor's discretion. (See SMM G20–S2.)

Part. 2&3/4,5&7
O/a dims.   8·050
−walls $^2/$0·255. 0·510
7·540

$\dfrac{7·54}{}$	100 × 50 mm Ditto	
$2/\dfrac{1·85}{2·46}$		(part 1&2/3
		(·· 4/5
		(& ·· 5/6&7

SMM G20.7.0.1.0 – S2

There is no need to describe the cill to partition 2 & 3/4, 5 & 7 as exceeding 6 m long as, when adjustment is made for the door opening, no length of cill will exceed 6 m.

Length of studs
2·475
− Hd. & cill $^2/$0·050: 0·100
2·375

Parts 1/2 & 6/7

$^2/2$ $/1·87$	75 × 50 mm Swn.	(Hd. & noggins
$^2/6$ $/2·38$	Tan. swd. wall	(studs
	or part. members.	

SMM G20.7.0.1.0

The noggins have been measured overall the studs and adjustment made later.

Stud Part. 5.

<u>Adj: for vert. studs</u>.

Noggins measured overall studs. (See previous note.)

2/6/0·05   <u>Ddt</u> 75 × 50 mm swn. Ton. swd. wall or part. members a.b.

SMM G20.7.0.1.0

<u>100 mm. thick parts.</u>

2/ 7·54	100 × 50 mm ditto
2/ 1·85	
2/ 2·46	
6·9 6/ 2·38	

head & noggins
(part. 2&3/4,5&7
(part. 1&2/3
("   4/5
(& " 5/6&7
(studs

6·
4·
3·
7·
5·
3

SMM G20.7.0.1.0

Although the head member of partition 2 & 3/4, 5 & 7 exceeds 6 m long it need not be in one length as it is fixed to the roof structure.

<u>Adj: for vert. studs</u>

Noggins measured overall studs. (See previous note.)

6·9 6/0·05   <u>Ddt</u> 100 × 50 mm ditto

SMM G20.7.0.1.0

6·
4·
3·
7·
5·
3

Name of Project.

date

Name 5

Stud Part. 6

Adj. for fixing end studs to walls.
75mm thick part.

2/2.38	Ddt 75 × 50 mm. swn. Tan. swd. wall or part. members	SMM G20.7.0.1.0

&

	Add 75 × 50 mm. ditto p α fxd. to blkwk. at 450 mm centres.	SMM G20.7.0.1.0 – S2

100mm thick part.

5/2.38	Ddt 100 × 50 mm ditto	SMM G20.7.0.1.0

&

	Add 100 × 50 ditto p α fxd. to blkwk. at 450mm centres.	SMM G20.7.0.1.0 – S2

100mm wide d.p.c.
part. 2&3/4,5&7 : 7·540
  ..   1&2/3   : 1·850
  "   4/5   = 2·460
  ..   5/6&7 : 2·460
                14·310

Name of Project.

Name 6   date

<u>Stud Part. 7.</u>

<u>D.p.c (Ctd)</u>

²⁄₁.87	*Pitch Polymer d.p.c. (75mm part.*
0.08	*width ⋞ 225 m.m.*
14.31	*horiz. single layer, (100mm ··*
0.10	*150 mm laps bedded in*
	g.m (1:1:6) (meas. nett)

SMM F30.2.1.3.0

<u>Top restraints.</u>

centres of rf. jsts. 0.400
− ²⁄₂/0.050 · 0.050
noggin length: 0.350

<u>Nr. of noggins</u>
part 1&2/3. 0.450)1.850
                    4 + 1 end= 5
part 4/5. 0.450)2.460
                5+1 rem+1end= 7
part 5/6&7 as part 4/5   = 7
                                    19

¹⁹⁄0.35	*50× 50 mm Swn. Tan. swd.*
	individual support.

Where partitions run parallel with the roof joists and falls between two joists, bearers are required to fix the head of the partition.

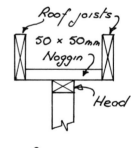

Roof joists
50 × 50mm
Noggin
Head

<u>SECTION.</u>

SMM G20.13.0.1.0

Stud Parts 8.

Adj. for door openings.

<u>Drs. in parts 1/2</u>
         ·· 6/7
      <u>Cills</u>

2/0·75   <u>Ddt</u> 75 × 50mm swn. Tan. swd. wall or part. member p. & fxd. to conc. a.b.   SMM G20.7.0.1.0 – S2

<u>Drs. in parts 3/5</u>
      ·· 4/5

2/0·90   <u>Ddt</u> 100 × 50mm ditto.   SMM G20.7.0.1.0 – S2

Noggin   0·750
   – stud   <u>0·050</u>
          <u>0·700</u>

2/0·70
2/2·10   <u>Ddt</u> 75 × 50 mm swn (noggin Tan. swd. wall or (stud part. member a.b.   SMM G20.7.0.1.0

Noggin   0·900
   – stud   <u>0·050</u>
          <u>0·850</u>

2/0·85
2/2·10   <u>Ddt</u> 100 × 50mm ditto (noggin (stud   SMM G20.7.0.1.0

<u>Stud Parts. 9</u>

<u>Adj</u>. dr. opgs. (Ctd)

2/0.75   <u>Add</u> 75×50 mm swn. (hd. of opg.   SMM G20.7.0.1.0
           Tan. swd. wall or
           part. member a.b.

2/0.90   <u>Add</u> 100×50 mm ditto ("   SMM G20.7.0.1.0

<u>dpc</u>.

2/0.75   <u>Ddt</u> d.p.c. ≤ 225 wide   SMM F30.2.1.3.0
  0.08   a.b.
2/0.90
  0.10

<u>To Take</u>

Finishings on stud
partitions to take
with Internal Finishings
section.

This note will make sure that
the plasterboard and
finishings applied to the studding
is not forgotten.

Name of Project.

date

Name 9

Roofs

Introduction

A roof, whether flat or pitched, is divided into the following sections for measurement purposes:

1. roof constructions
2. roof coverings
3. treatment at eaves
4. rainwater goods
5. roof lights and adjustments
6. adjustments for chimney stacks

Only the first four sections have been considered in the following example, as in practice the last two may be measured by another member of the taking-off team, as has been assumed here.

As mentioned in Chapter 6, it may be necessary to measure with this section any brick gable and party walls above roof plate level if the format of the bill is elemental. This brickwork would be measured with the first section, roof construction; however, the more traditional format has been adopted in the measurement of the following examples.

Obviously, it is preferable to have a plan of the roof, section through the roof and building, together with full specification but, in the absence of such, the roof layout may be superimposed on the upper floor plans and the overall dimensions calculated.

Approach

Divide the project into individual buildings and consider each building according to the style of roof, e.g. flat or pitched roofs, and then according to the form of construction, e.g. timber flat roof or concrete flat roof. Each roof is then further divided into the sections previously listed in the introduction and the constituent parts measured as shown in the taking-off lists.

Generally

Measure each individual roof including projections over all openings, chimney breasts, roof lights, etc., which are measured separately together with any adjustments. For measurement purposes, there are two classes of pitched roof:

1. Roofs which have a constant or same pitch on all roof slopes.
2. Roofs which do not have a constant pitch on all roof slopes.

In the following examples, constant pitched roofs have been used. For non-constant pitched roofs the same rules of measurement apply but each roof slope must be measured separately.

The plan of a rectangular constant pitched roof will show the shape of the roof and have the following geometric properties, as shown in Figures 9.1 and 9.2.

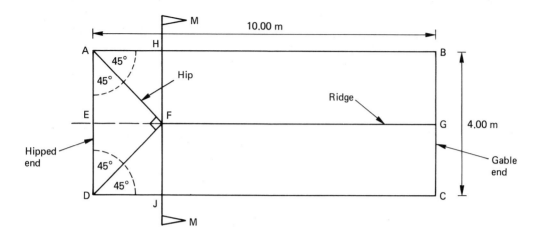

Figure 9.1   *Plan*

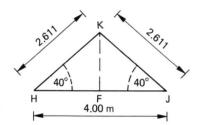

Figure 9.2   *Section M–M*

1. Shape of roof – pitched with one end hipped and other end gabled.
2. $\angle AFD = 90°$.
3. $\angle EAF = \angle EDF = 45°$.
4. $HF = EF = JF = $ Half span.

Using the above information, the total area of coverings can be calculated by dividing the roof into three individual areas – two trapeziums, which have the same area, plus the triangles at the hip, e.g.,

$$\text{Trapeziums} = 2.\left\{ \left(\frac{10+10-2}{2} \right) \times 2.611 \right\} = 46.998$$

  plus

$$\text{Triangles} = \frac{4 \times 2.611}{2} \qquad\qquad = 5.222$$

$$\underline{\underline{52.220\,\text{m}^2}}$$

However, because the roof is of constant pitch, the total area of coverings can be calculated by measuring overall, ignoring the hipped end, e.g. $2.(10 \times 2.611) = $ **52.220 square metres**. As the roof is of constant pitch, half the area of the hip end covering is equal to the additional area measured if one roof slope is measured as a rectangle. Similarly, when measuring the length of timber required for the common and jack rafters to the main roof slopes and hipped end, the combined length of each pair of jack rafters is nominally equivalent to the length of one common rafter. From Figures 9.3(a) and (b) it can be seen that: $a + a = $ common rafter.

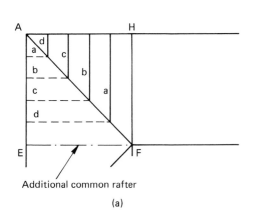

Additional common rafter

(a)

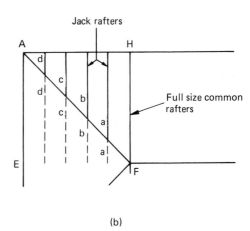

(b)

Figure 9.3 (a)   *Part plan at hipped end of roof (1)*

(b)   *Part plan at hipped end of roof (2)*

However, one additional common rafter must be measured to the apex of each hip to ensure that the required span between the last two jack rafters is not exceeded, see Figure 9.3.

Consider the division of the building outline into main roof and projections and in all cases in Figures 9.4–9.6 and the area of roof coverings will be the same, viz $2.(1 \times rs) + 2.(pl \times rs)$. As a result it can be seen that the shape of constant pitched roofs, with the inclusion of hips or valleys, has little effect on the measurement of roof coverings or common rafters.

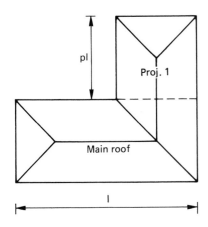

Figure 9.4   *Roof plan*

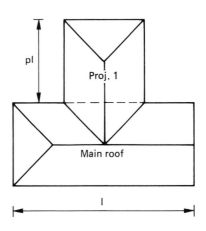

Figure 9.5   *Roof plan*

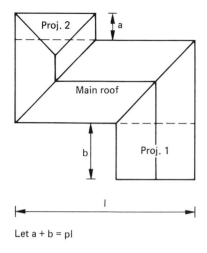

Let a + b = pl

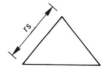

Figure 9.6   *Roof plan*   Figure 9.7   *Section*

Let length of roof slope in all cases be 'rs', see Figure 9.7.

Structural timbers – generally

Structural timber is measured in linear metres and cross-sectional dimensions are given in the description. Allowance must be made in the measurement for any joints required, e.g. halved joints or housed joints, etc. Structural timber is converted by sawing timbers into standard sizes known as nominal sizes. If a planed finish or wrot finish is required, then a finishing process is carried out which reduces the nominal size of the timber by approximately 3 mm on each finished face. As a result, a piece of timber planed all round having a nominal size of 50×100 mm will in fact be 44×94 mm which is known as the actual or finished size. Usually, sizes given on drawings and in bills of quantities are nominal but should finished sizes be required then the finished or actual size is given and indicated as such, see SMM G20–D1.

   If a section of timber is required to be in one length and exceeds 6 m long, then this is a requirement which must be stated in the description, see SMM G20.6.0.0.1. If a piece of timber is to be fixed in a specified way then this also needs to be stated in the description, otherwise the fixing will be left to the discretion of the contractor, see SMM G20–S2.

Rafters

In order to measure the amount of timber required for common and jack rafters it is necessary to:

1.   Calculate length of common rafter, see Figure 9.8.
     (a)   by Pythagoras's theorem. $L = \sqrt{S^2 + R^2}$
     (b)   by Trigonometry. $L = S \div \cos 40°$
           Calculation will give a dimension along the centreline of the rafter to which should be added an allowance for the raking and cutting at the ridge.
     (c)   By scaling. This method should only be used as a last resort.

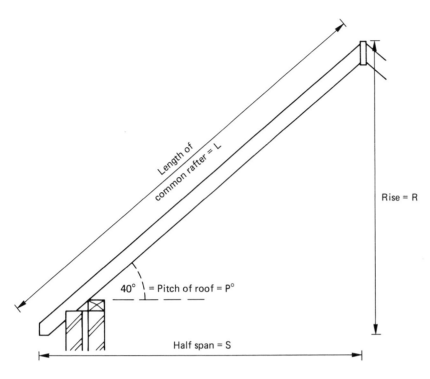

Figure 9.8 *Half section through roof*

2. Calculate number of common rafters.

   (a)   for any style of roof (not hipped).

      (i)   calculate the distance between the centre line of the first and last common rafters to one roof slope.

      (ii)   divide the distance by the centres between the rafters (this gives the number of gaps between the rafters).

      (iii)   add one rafter if there is a remainder (this ensures that the centres between rafters are not exceeded).

      (iv)   add one to the end rafter which closes the last gap.

    For example, let the centres of rafters be 450 mm.

   (i)   Calculate distance between centres as follows:

Overall dimension		2.810
less		
Gable end		
wall	0.255	
gap	0.025	
Centre of last rafter $\frac{1}{2}$/0.050	0.025	
	2/ 0.305 =	0.610
Distance between centres of first and		
last rafters		2.200

(ii)   Divide by centres   0.450)2.200 = 4
(iii)   Plus one for any remainder 0.400 = 1
(iv)   Plus one for end rafter   = 1
                                            —
                                            6 Nr

(b)   For hipped roofs:
    (i)   Calculate the extreme distance between hip rafters.
    (ii)   Divide the extreme distance by the centres between rafters.
    (iii)   Add one rafter if there is a remainder.
    (iv)   Deduct one to give the number of rafters between hip rafters.

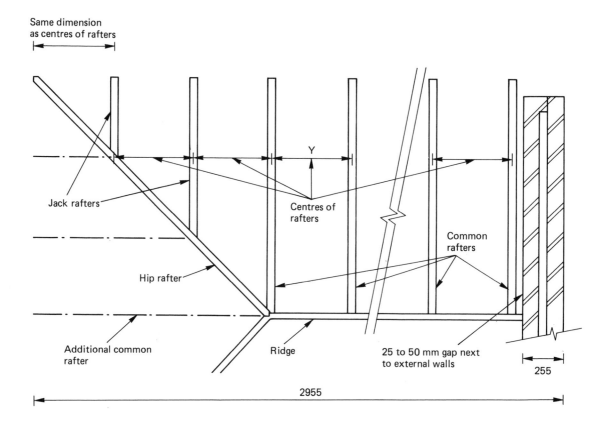

Figure 9.9   *Plan*

Ceiling joists

The number of ceiling joists is calculated in the same way as common rafters in (a) above but, depending on the roof overhang of the external walls, it is usual to have the same number of ceiling joists as common rafters.

Hip and valley rafters

It is necessary to find the length of hip and/or valley rafters as the plans and sections never show their true length.

1. By calculation using Pythagoras's theorem, see Figure 9.10.

$$\text{Hip} = \sqrt{\tfrac{1}{2}\ \text{span}^2 + \text{common rafter}^2}$$
$$= \sqrt{AE^2 + EK^2}$$

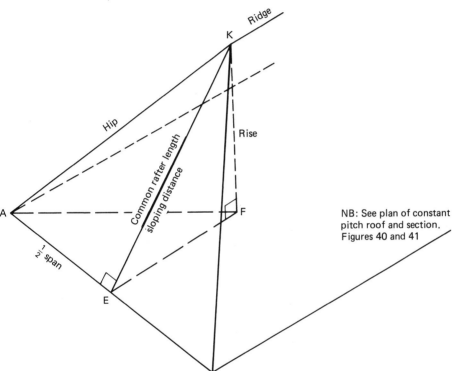

NB: See plan of constant
pitch roof and section.
Figures 40 and 41

Figure 9.10 *Isometric sketch of roof*

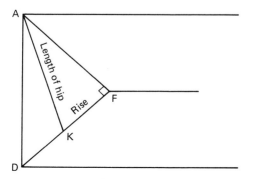

Figure 9.11 *Plan*

2. By scaling, see Figure 9.11.

    (a)   use plan of constant pitch roof
    (b)   scale on line of one hip the rise of the roof FK
    (c)   join points A to K to form a right-angle triangle. Scale AK = length of hip.

Coverings

The measurement of coverings, usually slates or tiles, follows the process previously described in this chapter, e.g. for constant pitch roofs, ignoring hips and valleys, measure the area of one roof slab overall and then multiply by two for the other slope. The dimensions to calculate the area must be taken to the centre line of the eaves gutters. The item of roof coverings includes all battening and underlay, see SMM H60.1–C1(a), and these must be included in the description. Following the covering, items required for perimeter work are measured usually in the order shown in SMM H60.3 to 9.

Eaves

Fascia, eaves and verge soffit boarding if less than 300 mm wide are all measured in linear metres. Greater widths may be measured in square metres. All boarding is measured to its extreme length to allow for mitred or raking cut joints.

    Decoration of the boarding is taken at the same time and it is usual to measure priming to the back of externally fixed joinery before fixing, see Figure 9.12. Each board will have an item for priming the back measured in accordance with SMM M60.1.0.0.4. The painting of the external face painted is grouped in with adjacent boards, e.g. fascia and soffit boards.

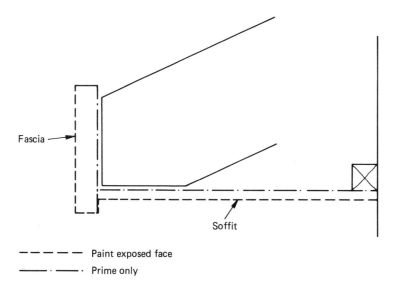

Fascia

Soffit

– – – – –   Paint exposed face
——— · ——— ·   Prime only

Figure 9.12   *Section through eaves*

    To comply with Building Regulations it is necessary to ventilate the roof and this should be measured in this section.

Rainwater goods

Rainwater pipes and eaves gutters are measured in linear metres over all fittings in accordance with SMM R.10.1 or .10 and then the fittings are measured as extra over the pipe or gutter in which they occur. If the rainwater goods are not self finished, then decoration will have to be measured. Painting to pipes is measured in linear metres as general isolated surfaces, stating the girth (as SMM M60.1.0.2). The girth of some larger pipes may exceed 300 mm, in which case it should be measured in square metres as general surfaces.

Painting to eaves gutters is measured in linear metres as SMM M60.8 and care must be taken to ensure that the decoration of the gutterwork is for the inside and the outside.

Flow chart for pitched roof

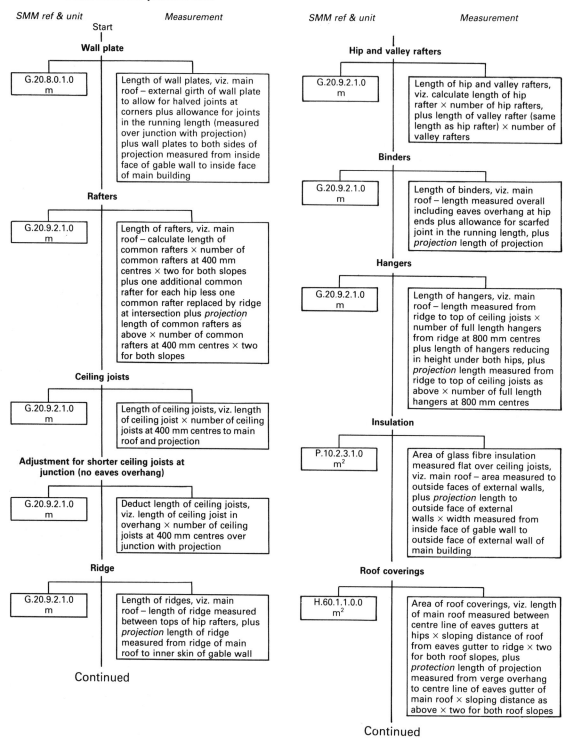

SMM ref & unit   Measurement

Start

Wall plate

| G.20.8.0.1.0 m | Length of wall plates, viz. main roof – external girth of wall plate to allow for halved joints at corners plus allowance for joints in the running length (measured over junction with projection) plus wall plates to both sides of projection measured from inside face of gable wall to inside face of main building |

Rafters

| G.20.9.2.1.0 m | Length of rafters, viz. main roof – calculate length of common rafters × number of common rafters at 400 mm centres × two for both slopes plus one additional common rafter for each hip less one common rafter replaced by ridge at intersection plus *projection* length of common rafters as above × number of common rafters at 400 mm centres × two for both slopes |

Ceiling joists

| G.20.9.2.1.0 m | Length of ceiling joists, viz. length of ceiling joist × number of ceiling joists at 400 mm centres to main roof and projection |

Adjustment for shorter ceiling joists at junction (no eaves overhang)

| G.20.9.2.1.0 m | Deduct length of ceiling joists, viz. length of ceiling joist in overhang × number of ceiling joists at 400 mm centres over junction with projection |

Ridge

| G.20.9.2.1.0 m | Length of ridges, viz. main roof – length of ridge measured between tops of hip rafters, plus *projection* length of ridge measured from ridge of main roof to inner skin of gable wall |

Continued

SMM ref & unit   Measurement

Hip and valley rafters

| G.20.9.2.1.0 m | Length of hip and valley rafters, viz. calculate length of hip rafter × number of hip rafters, plus length of valley rafter (same length as hip rafter) × number of valley rafters |

Binders

| G.20.9.2.1.0 m | Length of binders, viz. main roof – length measured overall including eaves overhang at hip ends plus allowance for scarfed joint in the running length, plus *projection* length of projection |

Hangers

| G.20.9.2.1.0 m | Length of hangers, viz. main roof – length measured from ridge to top of ceiling joists × number of full length hangers from ridge at 800 mm centres plus length of hangers reducing in height under both hips, plus *projection* length measured from ridge to top of ceiling joists as above × number of full length hangers at 800 mm centres |

Insulation

| P.10.2.3.1.0 m² | Area of glass fibre insulation measured flat over ceiling joists, viz. main roof – area measured to outside faces of external walls, plus *projection* length to outside face of external walls × width measured from inside face of gable wall to outside face of external wall of main building |

Roof coverings

| H.60.1.1.0.0 m² | Area of roof coverings, viz. length of main roof measured between centre line of eaves gutters at hips × sloping distance of roof from eaves gutter to ridge × two for both roof slopes, plus *protection* length of projection measured from verge overhang to centre line of eaves gutter of main roof × sloping distance as above × two for both roof slopes |

Continued

Flow chart for pitched roof – continued

SMM ref & unit	Measurement	SMM ref & unit	Measurement

Continued

Eaves

H.60.4.0.0.0 m	Length of eaves, viz. girth of main roof measured on centre line of eaves gutter, less width of projection measured between centre line of eaves gutters at junction of main roof and projection, plus *projection* length of projection roof as above × two for both roof slopes

Verges

H.60.5.0.0.0 m	Length of verges at gable end, viz. sloping distance of roof as above × two for both roof slopes

Ridge

H.60.6.0.0.0 m	Length of ridge tiling, viz. main roof – length of ridge tiles measured between tops of hip rafters, plus *projection* length of ridge tiles measured from ridge of main roof to verge overhang at gable end

Eaves ventilation

H.60.10.1.1.0 Measurement in linear metres. This differs from the requirements of SMM and must be recorded in the preliminary bill	Length of ventilation units of eaves, viz. same length as eaves measurement used above

Hip tiles

H.60.7.0.0.0 m	Length of hip tiling, viz. calculate length of hip tiling × number of hips

Valley tiles

H.60.9.0.0.0 m	Length of valley tiling, viz. same length as calculated for one hip above × number of valleys

Continued

Fascia board

G.20.15.3.2.0 m	Length of fascia boards, viz. external girth of fascia for complete building less gable end, plus adjustment for two internal mitres at junction of main roof and projection

Prime only back of fascia board

M.60.1.0.2.4. m	Length of painting back of fascia board, viz. same length as fascia boards above

Soffit board

G.20.16.3.2.0 m	Length of soffit boarding, viz. external girth of soffit boarding for complete building less gable end plus adjustment for two internal splay cut angles at junction of main roof and projection

Prime only back of soffit board

M.60.1.0.2.4 m	Length of painting back of soffit boarding, viz. same length as soffit boards above

Sawn softwood bearer

G.20.13.0.1.0 m	Length of bearer supporting soffit boards, viz. external girth of bearer fixed to external wall of complete building less gable wall

Boxed ends to eaves

G.20.18.0.1.0 nr.	Number of boxed ends to eaves at gable end

Prime only back of boxed ends

M.60.1.0.3.4 nr.	Number of boxed ends to eaves

Painting fascia and eaves boarding

M.60.1.0.1.0 m²	Area of external general surfaces of paint, viz. length of fascia boards above × girth of exposed face of fascia and soffit boarding plus area of boxed ends

Continued

Flow chart for pitched roof – continued

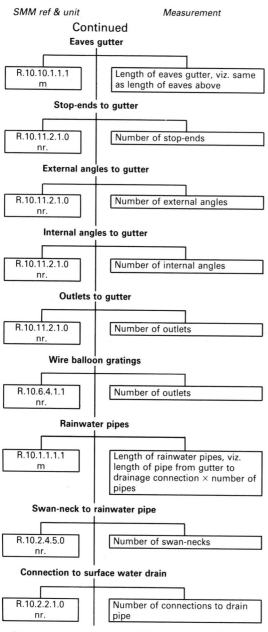

SMM ref & unit   Measurement

Continued

Eaves gutter

| R.10.10.1.1.1 m | Length of eaves gutter, viz. same as length of eaves above |

Stop-ends to gutter

| R.10.11.2.1.0 nr. | Number of stop-ends |

External angles to gutter

| R.10.11.2.1.0 nr. | Number of external angles |

Internal angles to gutter

| R.10.11.2.1.0 nr. | Number of internal angles |

Outlets to gutter

| R.10.11.2.1.0 nr. | Number of outlets |

Wire balloon gratings

| R.10.6.4.1.1 nr. | Number of outlets |

Rainwater pipes

| R.10.1.1.1.1 m | Length of rainwater pipes, viz. length of pipe from gutter to drainage connection × number of pipes |

Swan-neck to rainwater pipe

| R.10.2.4.5.0 nr. | Number of swan-necks |

Connection to surface water drain

| R.10.2.2.1.0 nr. | Number of connections to drain pipe |

Check drawings and taking-off list for outstanding items

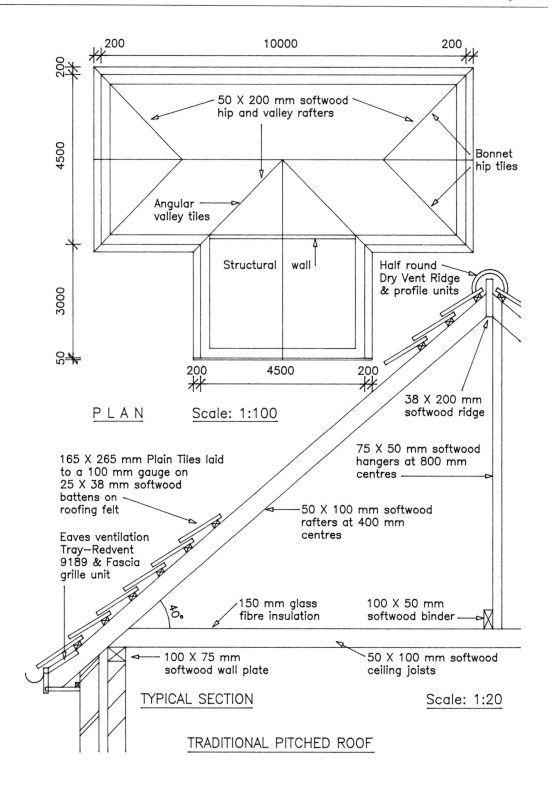

200 10000 200

200

4500

50 X 200 mm softwood
hip and valley rafters

Bonnet
hip tiles

Angular
valley tiles

3000

Structural wall

Half round
Dry Vent Ridge
& profile units

50

200 4500 200

P L A N   Scale: 1:100

38 X 200 mm
softwood ridge

165 X 265 mm Plain Tiles laid
to a 100 mm gauge on
25 X 38 mm softwood
battens on
roofing felt

75 X 50 mm softwood
hangers at 800 mm
centres

Eaves ventilation
Tray—Redvent
9189 & Fascia
grille unit

50 X 100 mm softwood
rafters at 400 mm
centres

40°

150 mm glass
fibre insulation

100 X 50 mm
softwood binder

100 X 75 mm
softwood wall plate

50 X 100 mm softwood
ceiling joists

TYPICAL SECTION   Scale: 1:20

TRADITIONAL PITCHED ROOF

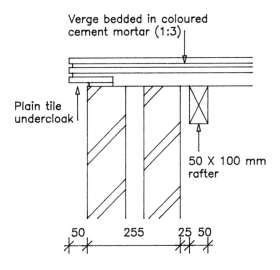

Verge bedded in coloured cement mortar (1:3)

Plain tile undercloak

50 X 100 mm rafter

50 255 25 50

SECTION THROUGH VERGE   Scale: 1:10

SPECIFICATION
1. All sawn softwood to be stress graded timber to CP 112 grade SS group S1 and pressure impregnated with Tanalith C. dry salt net retention 5.3 kg / m³.
2. Roof tiling to be Redland Rosemary plain tiles — colour Medium Mixed Brindle laid to a 65 mm lap, each tile in every fifth course twice nailed with 38 mm X 12g aluminium alloy nails.
3. Ventilation:
   a) Eaves — Redland Red Vent eaves ventilator with fascia grille.
   b) Ridge — Redland Dry Vent ridge system with plastic air vents both sides.
4. All wrot softwood to be primed all round and painted — 2 undercoats and 1 top coat of paint on exposed faces.

TRADITIONAL PITCHED ROOF

Pitched Roof 1

Taking Off List.

Construction

1. Wall plates
2. Rafters
3. Ceiling joists
4. Ridge
5. Hip & Valley rafters
6. Binders
7. Hangers
8. Insulation

Coverings

1. Tiling including battens & felt.
2. Eaves   ..   ventilators
3. Verges
4. Ridges including ventilators
5. Hips
6. Valleys.

Eaves

1. Fascia including priming backs
2. Soffit   ..   ..
3. Boxed ends ..   ..   ..
4. Decoration

Rainwater Goods

1. Gutters including fittings
2. Rainwater pipes "   "

Drawing Number.

Name of Project

date

Name 1

Page Nrs.

1	8	15	22
2	9	16	23
3	10	17	24
4	11	18	25
5	12	19	26
6	13	20	27
7	14	21	28

Pitched Roof 2

Construction.

Preamble Note.
All structural timber is to be stress graded sawn softwood to C.P. 112 grade SS. group S₁, and to be pressure impregnated with 'Tanolith C' dry salt net retention 5.3 kg./m³

To comply with the requirements of SMM G20–S1 and S4 a preamble note is inserted in the dimensions. This also saves having to repeat the full description for each structural softwood item.

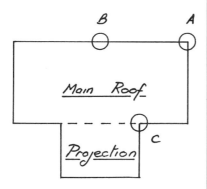

As the pitch of the roof is constant, the shape has little effect on the method of measurement. Therefore for measurement purposes it is split into main roof and projection. (See diagram.)

Wall plates

Main roof

	10·000
	4·500
²⁄	14·500
ext. girth of bwk.	= 29·000
– wall passings ²/4/²/⅟₂/0·255	2·040
int. girth of bwk.	26·960
+ plate passings ²/4/²/⅟₂/0·100	0·800
ext. girth of wall plate	27·760
+ jts. in the running length where plates ⅋ 6·00m long ²/0·150	0·300
	28·060

The wall plate is measured on the external girth to allow for halved joints at the corners.

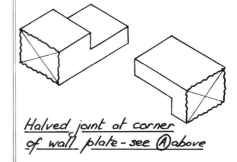

Halved joint at corner of wall plate – see Ⓐabove

Pitched   Roof_3

Constr. (Ctd)
Wall plate(Ctd)

A wall plate is required to support the main roof ceiling joists across the junction of the main roof and projection. The wall plate will run on the top of the structural partition which is shown on the drawing. If there were no partition shown then a beam would have to be measured.

The wall plate to the two long sides exceeds 6.00 m long. It is not necessary to have it in one continuous length as it is supported on a wall, therefore an allowance is made for a halved joint. (See SMM G20.8.0.0.1.)

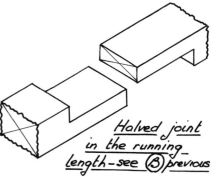

Halved joint
in the running
length-see Ⓑ previous

SMM G20.8.0.1.0

The wall plate to the projection is measured over the main roof plate and to the inside face of the gable wall, thus allowing for a halved joint at the junction.

2/	28·06	100 × 75mm Swn. Tan. swd plate bedded in (proj. g. m (1:1:6)
	3·00	

Name of Project.

date   Name of Project.

Name 3

Pitched Roof 4.

Constr (Ctd)
Wall plate (Ctd)

<u>NB.</u> There is no wall
plate at the
gable end

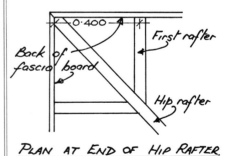

Main roof plate

Projection plate

<u>Halved joint at
junction</u> - see Ⓒ previous.

The main roof is essentially the same as a roof hipped at both ends. Therefore the rafters will be set out from the ends of the hip rafters. The overall dimension is calculated from the back of the fascia at the hips.

<u>Rafters</u>
<u>Numbers</u>

<u>Main roof</u>.
 o/a bk. dims.   10·000
+ overhang   0·200
− fascia bd.   0·025
$^2/$ 0·175 = 0·350
0·400) 10·350
= 25 + 1 rem
− 1 end
= **25 Nr.**

Back of fascia board

First rafter

Hip rafter

⊢ 0·400 ⊣

<u>PLAN AT END OF HIP RAFTER.</u>

An allowance of one is made for any remainder to ensure that

Pitched Roof 5.

<u>Constr. (Ctd)</u>
<u>Rafters(Ctd)</u>

the centres are always less than the design centres of the rafters.

The amount of timber measured for rafters will be sufficient to produce the structure to cover the main roof irrespective of its shape.

The projection is treated as a roof structure on its own. See drawing for position of the first rafter next main roof.

<u>Numbers (Ctd)</u>

<u>Projection</u>

o/a. bk. dims.		3·000		
– junction with main rf.				
overhang	0·200			
– fascia	0·025	:	0·175	
			2·825	
– gable wall	0·255			
gap	0·025			
¢ rafter ½/0·050 0·025		:	0·305	
¢ 1st to last rafters			2·520	

$$0·400 \overline{)2·520}$$
$$: 6 + 1rem + 1end$$
$$= 8 Nr.$$

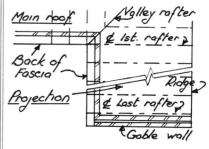

PART PLAN OF PROJECTION

<u>Length of common rafters</u>

half span bk. dims. ½/4·500:		2·250	
+ eaves overhang:	0·200		
– fascia bd.	·0·025 :	0·175	
		2·425	
– ridge	½/0·038	0·019	
		2·406	

by trigonometry length = $\dfrac{half\ span}{cos\ pitch}$

$= \dfrac{2·406}{cos\ 40°}$

C/f = 3·141

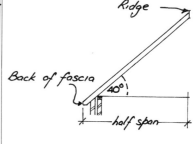

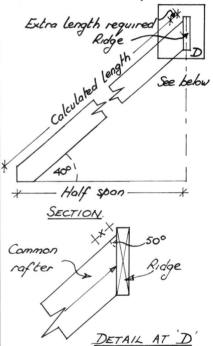

Pitched Roof 6

Constr. (Ctd)

__Length of common rafters (Ctd)__

$$b/f = 3.141$$

$$\text{extra length} \times$$

$$= \frac{0.050}{\text{Tan } 50°}$$

$$= \frac{0.042}{3.183}$$

The shape of common rafters is complex for calculation purposes. Because there are so many rafters in a roof it is necessary to calculate the length of timber required for each rafter accurately.

The calculation for this additional length of timber shows that there could be an undermeasure. In practice an allowance of half the depth of the rafter is added to the calculated length to cover the undermeasure.

Name 6   date   Name of Project

Pitched Roof_7_

Constr. (Ctd)
Rafters (Ctd)

2/ 25/ 8/ 3·18	50 × 100 mm Swn. 'Tan.' swd. rf. members (Add. rft. at hip pitched. (see below.
3·18	

SMM G20.9.2.1.0

25 and 8 are the numbers of rafters and timesing by 2 is for both sides of the roof.

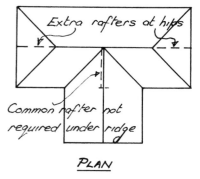

Extra rafters at hips

Common rafter not required under ridge

PLAN

Additional rafters
hips	:	2 Nr
− rafter not required	:	1 Nr
	:	1 Nr

Usually there will be one ceiling joist for every pair of common rafters, but if a hip ended roof has wide overhangs at eaves then the number of ceiling joists should be calculated.

Ceiling joists.
Nr.
Main roof
o/a bwk. dims	10·000
− wall	0·255
− gap	0·025
− jst. ½/0·050·0·025	
²/0·305 = 0·610
0·400) 9·390

= 23 + 1 rem + 1 end
= 25 Nr.

Projection
Nr
ext. bwk. dims	3·000
+ ext wall main rf.	0·255
	3·255
− part.	0·103
− gable wall	0·255
− gaps ²/0·025·0·050
− jsts 2/½/0·050·0·050·0·458
0·400)2·797

= 6 + 1 rem + 1 end = 8 Nr.

Name 8 date Name of Project.

Pitched Roof 8

Constr (Ctd.)
Clg. jsts. (Ctd)

8²⁵/4·50   50 × 100 mm. Swn. 'Tan.' swd. rf. members pitched

SMM G20.9.2.1.0

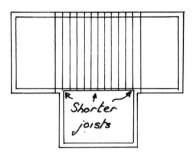

PLAN

The ceiling joists at the junction with the projection are shorter than the general joists as they finish on the partition.

Adj. of clg. jsts. at junction with proj.

wall	0·255
−partition	0·100
	0·155

	Nr.
proj. width	4·500
+eaves overhang	0·200
−fascia	·0·025
−jst ½/0·050·0·025 ÷ 0·050	
+ ²/0·150·0·300	
0·400) 4·800	
÷	12−1 end
÷	11 Nr.

The number is calculated using the design centres of the joists, but this time one is deducted from the number to give the number of shorter joists, leaving the two extreme joists the full length.

11/0·16   Ddt   50 × 100mm ditto

SMM G20.9.2.1.0

Name of Project.

date

Name 9

Pitched Roof 9.

Constr. (Ctd)
Ridge

Main roof		10.000
− hips	2½/4.500	4.500
		5.500

Projection		3.000
+ main rf. ½/4.500 :	2.250	
		5.250
− gable wall :	0.255	
gap :	0.025	
ridge ½/0.038 : 0.019	0.299	0.299
		4.951

5.50	
4.95	

38 × 200mm Swn. Tan. (main rf.
swd. rf. members (proj.
pitched.

SMM G20.9.2.1.0

Hip & valley rafters.

$$\text{hip rafter} = \sqrt{\text{half span}^2 + \text{rafter}^2}$$
$$= \sqrt{2.425^2 + 3.183^2}$$
$$= \underline{4.002}$$

VIEW OF HIP.

2⁴/4.00	

50 × 200 mm Ditto

SMM G20.9.2.1.0

(valley rafters dotted on.

Hip and valley rafters are the same length.

<u>Pitched Roof 10</u>

<u>Constr. (Ctd)</u>
Binders

10·55	2·90	

<u>Main roof</u> | 10·000
+ eaves overhang ²/0·175·0·350 | 0·350
| 10·350
+scarf jt. ²/0·100: | 0·200
| 10·550

<u>Projection</u> | 3·000
- part. | 0·100
| 2·900

PLAN

It would be impractical to provide timber 10 m long so an allowance is made for a scarfed joint.

d = 100 mm

2d = 200 mm

SCARF JOINT.

SMM G20.9.2.1.0

50×100 mm Swn. (main rf.
'Tan'. swd. rf. (proj.
members pitched.

<u>Hangers.</u>
<u>Length</u> : Tan 40° × 2·250
: 1·888 m

<u>Nr. full length.</u>
main rf. ridge length | 5·500
0·800) 5·500
= 6 + 1 rem + 1 end
= 8 Nr

40°
2·250
length of hanger

Pitched Roof 11

Name // date   Name of Project.

Constr. (Ctd)
Hangers (Ctd)
Nr. full length

projection   binder length = 2·900

= 0·800) 2·900

: 3 + 1rem + 1end

= 5 Nr.

Ridge

Full length hanger

6

a   40

*800 * 800 *   Binder

—— 2·425 ——

SECTION THROUGH HIP.

Hangers are still required to
support the binder under the
hipped ends.

Length under hip

hanger 'a'   2·425

      − 2/0·800 = 1·600

                    0·825

= 0·825 × Tan. 40°

= 0·692

hanger 'b'   2·425

        −   0·800

                    1·625

= 1·625 × Tan. 40°

= 1·364

5 8/1·89	
2/0·69	
2/1·36	

50 × 75 mm. Swn. (full length
'Tan' swd. roof   (a-hip
members pitched. (b-hip.

SMM G20.9.2.1.0

Hanger

Length of
hanger calculated
to bottom of
joist.

Hanger starts
at top of joist.

Ceiling joist

DETAIL

Adj. for clg. jsts.

2 5 8/0·10	
2	

Ddt 50 × 75 mm ditto.

SMM G20.9.2.1.0

Pitched Roof 12

<u>Constr. (Ctd)</u>
<u>Insulation.</u>

	Projection	3·000
	- gable wall	0·255
		2·745

10·00	150 mm Glass fibre (main rf.
4·50	insulation quilts
4·50	laid bet. members (proj.
2·75	at 400 mm. centres,
	butt joints horiz.

SMM P10.2.3.1.0

<u>Coverings.</u>

	Main roof	10·000
	+ overhang	0·200
	+ over gutter	0·050
	hips. ²⁄0·250	0·500
		10·500
	Projection	3·000
	-overhang etc. main rf. ab.	0·250
		2·750
	+ verge	0·050
		2·800

Slpg. length

half span: ½⁄4·500 :	2·250
+ overhang etc a.b. :	0·250
	2·500
slpg. length :	$\frac{2·500}{\cos 40°}$
	= 3·264

SECTION
(sloping length, 40°, 2·500)

Pitched Roof 13.

Cvgs. (Ctd)

2/10·50	165 × 265 mm 'Redland (main rf.
3·26	Rosemary' plain
2/2·80	tiles as spec. rf. (proj.
3·26	cvgs. 40° pitch

laid to a 65mm end lap
& half tile side lap, each
tile in every 5th cors.
twice nailed with 38mm ×
12g. aluminium alloy nails
to& inc. 32×19 mm swn. Tan.
swd. battens at 100mm
gauge. fxg. with 40mm × 12
g. nails inc. rfcd. bit.
rfg. felt underlay to
BS. 747 type 1F fxd. with
galvanised clout headed
nails with 150mm laps
at all passings.

SMM H60.1.1.0.0

For constant pitch roofs this
measurement will give the
correct area of tiling etc.
irrespective of the shape.

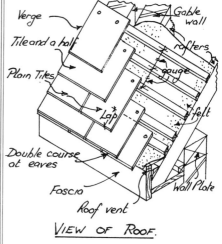

VIEW OF ROOF.

	Eaves
Main rf.	10·500
2/2·500 =	5·000
	2/15·500
	= 31·000
− junction	5·000
	26·000

Name of Project.

date

Name 13

Name

Name of Project.　Name 14　date

Pitched Roof 14.

Crgs. (Ctd)
Eaves (Ctd)

2/ 26·00 2·80	Eaves, double (main rf. course, each tile (proj. twice nailed.

&

'Redland Redvent' eaves
ventilator Type 9189
with fascia grille &
integral apron, fxg. to
swd. rafters at 400 mm
centres.

PRELIMINARY NOTE
State change of unit for
measurement of roof
ventilator.

Verges
proj.

2/ 3·26	Verges inc. tile & half & plain tile undercloak, bedded & ptd. in col. c.m. (1:3) each tile twice nailed.

SMM H60.4.0.0.0

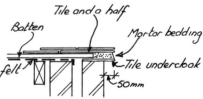

Double course
at eaves
Roof vent
felt
EAVES.

SMM H60.10.1.1.0
(unit altered to m)

The eaves ventilation system
specified here is a continuous
system and therefore has been
measured in linear metres
rather than numbered as
required by SMM. This change
must be recorded in the
Preliminary Bill.

Tile and a half
Batten   Mortar bedding
felt   Tile undercloak
50mm

SECTION THROUGH VERGE.
SMM H60.5.0.0.0

SMM H60–C2 states that
boundary work is deemed to
include undercloaks, bedding,
pointing etc. but SMM H60–S5
requires the method of forming
verges etc. to be stated.

<u>Pitched Roof 15.</u>

<u>Cvgs. (Ctd)</u>
<u>Ridge.</u>

<u>Proj</u>:   3·000
+ main rf. $\frac{1}{2}$/4·500 2·250
+ verge overhang : <u>0·050</u>
                          5·300

<u>5·50</u> <u>5·30</u>	Half round ridges (main rf. to match gen. tiling (proj. inc. 'Redland DryVent' ridge system, fixing with stainless steel nails & neoprene washers, plastic air vents both sides.	SMM H60.6.0.0.0

<u>Hips</u>

length : $\sqrt{\text{half span}^2 + \text{slpg. dist.}^2}$

: $\sqrt{2\cdot500^2 + 3\cdot264^2}$

: <u>4·111</u>

Length of hip tiles calculated by Pythagoras's theorem as the hip and valley rafters.

4/4·11	Bonnet hip tiles to match gen. tiling each tile nailed & bedded in col. c.m. (1:3)	SMM H60.7.0.0.0

<u>Valleys.</u>

2/4·11	Angular valley tiles to match course & bond with gen. tiling.	SMM H60.9.0.0.0

Name of Project.   *date*   *Name 15*

Pitched Roof 16.

<u>Eaves.</u>
<u>Fascia bd.</u>
10·000

main rf.	4·500	
proj.	3·000	2/ 7·500
		17·500

ext. girth of bwk = 35·000
− gable end   4·500
         30·500
+ passings 2/2/2/½/0·200 · 0·800
ext. face of fascia   31·300
+ mitre at int. ∠ 2/7/0·025 = 0·100
+ verge   2/0·050 = 0·100
             31·500

<u>To calculate passing timesing.</u>
    ext. ∠ = 4
    int. ∠ = 2
          2

31·50   25 × 125 mm Wrot. swd. fascia bd. width ≤ 300 mm

SMM G20.15.3.2.0

Timber is measured the length
necessary to cut the mitres
and angles etc.

&

Prime only gen. surf. of wood isol. surf. girth ≤ 300 mm applied on site prior to fxg.

SMM M60.1.0.2.4

<u>Soffit bd.</u>
girth of fascia bd.   31·500
− passing 2/4/2/½/0·025   0·200
girth of back of fascia   31·300
overhang   0·200
− fascia   0·025 = 0·175
       2/7/0·175
+ int. angle   2/7/0·175   0·700
              32·000

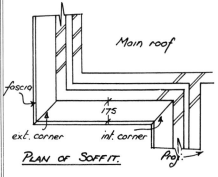

PLAN OF SOFFIT.

Name of Project.

date

Name 16

Pitched   Roof 17

Eaves (Ctd)
Soffit (Ctd)

32·00	19 × 175 mm *Wrot swd.* eaves soffit. bds. width ≤ 300 mm .	SMM G20.16.3.2.0
	&	
	Prime only gen. surf. of wood isol. surf. girth ≤ 300 mm a.b.	SMM M60.1.0.2.4

bwk. girth a.b.	30·500
+ *passings* 2/2/½/0·025	0·100
	30·600

30·60	25 × 50mm *Swn. swd.* individual supports plugged & fxd. to bwk. at 450mm centres.	SMM G20.13.0.1.0

Pitched Roof 18.

Eaves (Ctd)
Boxed ends.

2/1

19mm Wrot swd. spandril ends to eaves 200 × 300 mm overall. (Scaled

SMM G20.18.0.1.0

Boxing in the gap between the fascia and soffit boards and brickwork at the gable ends.

&

Prime only gen. surf. of wood, isol. area ≤ 0·50m² irrespective of girth applied on site prior to fxg.

SMM M60.1.0.3.4

Decoration.
girth of eaves

overhang	0·200
fascia	0·125
recess	0·025
	0·350

SECTION THROUGH EAVES.

31·50
0·35
2/ 0·20
0·30

K.P.S × ③ on gen. surf. of swd. girth > 300 mm

(boxed ends

external

SMM M60.1.0.1.0
The term KPS & ③ refers to knotting, priming and stopping the timber and then painting three coats of paint. The specification and number of undercoats etc. will be specified in the Preambles.

Name of Project.

date

Name 18

Name of Project.

<u>Pitched   Roof 19</u>

<u>Rainwater goods</u>
Gutters.

same dims. as
for eaves cors.

26·00	
2/ 2·80	

100mm Dıa. U.P.V.C. rain-
-water gutters to B.S.
4576 str. half round
with clipped neoprene
jts. & fascıa bkts. at.
1m centres screwed
to swd.

SMM R10.10.1.1.1

Gutters are measured on the
centre line and all fittings
are measured as extra over.

2

Extra over ditto for stop ends

SMM R10.11.2.1.0

4

E.o. ditto for ext. L's.

SMM R10.11.2.1.0

2

E.o. ditto for int. L's

SMM R10.11.2.1.0

3

E.o. ditto for outlet with
nozzle for 75mm dıa.
pipe.

SMM R10.11.2.1.0

Name of Project.

date

Name 19

Pitched Roof 20

Rainwater Gds. (Ctd.)
Gutters (Ctd)

3	Pipework ancillaries gal. wire balloon grating to fit 75mm dia. pipe.	SMM R10.6.4.1.1

Pipes
NB. Length assumed

3/5·00	75mm Dia. U.P.V.C. rain- -water pipes to B.S. 4576 str. with push fit socketted jts. & brackets at 2m. centres p&s to bwk.	SMM R10.1.1.1.1
3/1	E.o. ditto for swanneck 200 mm proj.	SMM R10.2.4.5.0

&

	E.o. ditto for caulking bush & cmt. jt. to stoneware drain.	SMM R10.2.2.1.0

Name of Project.

date

Name 20

Name

10

Trussed rafter roof

Introduction

The notes given in Chapter 9 also apply to this example with the following additional comments.

Information required

In addition to plans and sections of roof and specification, it is usual that the roof be designed by a timber engineering specialist who will provide a list of the component parts for inclusion in the bills of quantities.

Measurement

Roof trusses and truss rafters, together with mono-pitch trusses and gable ladders are numbered, see SMM G20.1 or 2. These items are usually manufactured off site and delivered ready for erection. Truss rafters replace the common rafters, ceiling joints, struts, etc., and are normally calculated and placed at the centres required for the roof tiling battens. Roof trusses support purlins which, in turn, support common rafters.

The length of any diagonal bracing is calculated in the same way as for hip rafters.

All metal clips, hangers, etc., are numbered (see SMM G20.20–25), but nail plates fixed on the trusses, etc., at works are included with the item of the roof trusses, etc.

The coverings in this example are interlocking roof tiles and differ in measurement from plain roof tiles in the following way:

1. The item for double course at eaves is not required.
2. A special tile is required for the left-hand verge of a roof, otherwise an unfinished edge would be seen. Sometimes a stainless steel cover plate is measured to both verges.

Flow chart for trussed rafter roof

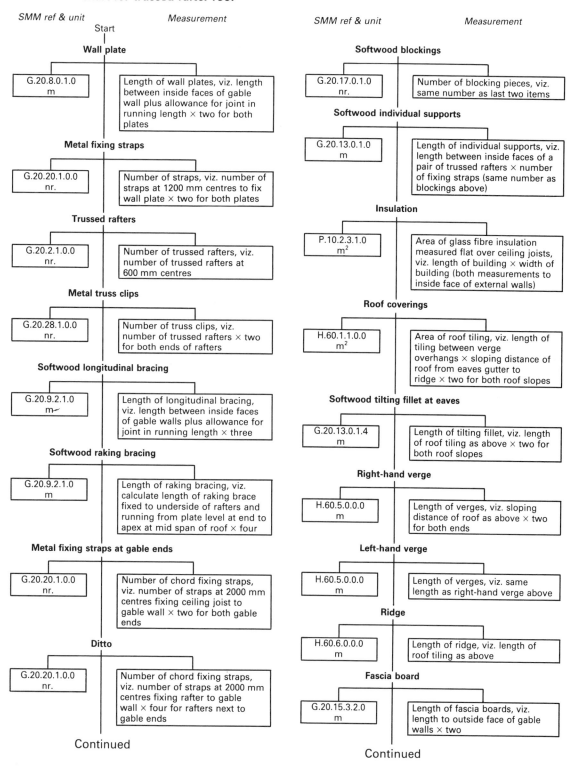

SMM ref & unit   Measurement

Start

Wall plate

G.20.8.0.1.0
m

Length of wall plates, viz. length between inside faces of gable wall plus allowance for joint in running length × two for both plates

Metal fixing straps

G.20.20.1.0.0
nr.

Number of straps, viz. number of straps at 1200 mm centres to fix wall plate × two for both plates

Trussed rafters

G.20.2.1.0.0
nr.

Number of trussed rafters, viz. number of trussed rafters at 600 mm centres

Metal truss clips

G.20.28.1.0.0
nr.

Number of truss clips, viz. number of trussed rafters × two for both ends of rafters

Softwood longitudinal bracing

G.20.9.2.1.0
m

Length of longitudinal bracing, viz. length between inside faces of gable walls plus allowance for joint in running length × three

Softwood raking bracing

G.20.9.2.1.0
m

Length of raking bracing, viz. calculate length of raking brace fixed to underside of rafters and running from plate level at end to apex at mid span of roof × four

Metal fixing straps at gable ends

G.20.20.1.0.0
nr.

Number of chord fixing straps, viz. number of straps at 2000 mm centres fixing ceiling joist to gable wall × two for both gable ends

Ditto

G.20.20.1.0.0
nr.

Number of chord fixing straps, viz. number of straps at 2000 mm centres fixing rafter to gable wall × four for rafters next to gable ends

Continued

SMM ref & unit   Measurement

Softwood blockings

G.20.17.0.1.0
nr.

Number of blocking pieces, viz. same number as last two items

Softwood individual supports

G.20.13.0.1.0
m

Length of individual supports, viz. length between inside faces of a pair of trussed rafters × number of fixing straps (same number as blockings above)

Insulation

P.10.2.3.1.0
m²

Area of glass fibre insulation measured flat over ceiling joists, viz. length of building × width of building (both measurements to inside face of external walls)

Roof coverings

H.60.1.1.0.0
m²

Area of roof tiling, viz. length of tiling between verge overhangs × sloping distance of roof from eaves gutter to ridge × two for both roof slopes

Softwood tilting fillet at eaves

G.20.13.0.1.4
m

Length of tilting fillet, viz. length of roof tiling as above × two for both roof slopes

Right-hand verge

H.60.5.0.0.0
m

Length of verges, viz. sloping distance of roof as above × two for both ends

Left-hand verge

H.60.5.0.0.0
m

Length of verges, viz. same length as right-hand verge above

Ridge

H.60.6.0.0.0
m

Length of ridge, viz. length of roof tiling as above

Fascia board

G.20.15.3.2.0
m

Length of fascia boards, viz. length to outside face of gable walls × two

Continued

Flow chart for trussed rafter roof – continued

SMM ref & unit	Measurement

Continued

Prime only back of fascia board

| M.60.1.0.2.4
m | Length of painting back of fascia board, viz. same length as fascia boards above |

Soffit board

| G.20.16.3.2.0
m | Length of soffit boards, viz. same length as fascia boards above |

Prime only back of soffit board

| M.60.1.0.2.4
m | Length of painting back of soffit boards, viz. same length as fascia boards above |

Sawn softwood bearer

| G.20.13.0.1.0
m | Length of bearer supporting soffit boards, viz. same length as fascia boards above |

Roof ventilation system

| To take note | Not measured in this example |

Boxed ends to eaves

| G.20.18.0.1.0
nr. | Number of boxed ends to eaves at gable ends |

Prime only back of boxed ends

| M.60.1.0.3.4
nr. | Number of boxed ends to eaves |

Painting fascia and eaves boarding

| M.60.1.0.1.0
m² | Area of external general surfaces of paint, viz. length of fascia boards above × girth of exposed face of fascia and soffit boarding, plus area of boxed ends |

Eaves gutter

| R.10.10.1.1.1
m | Length of eaves gutter, viz. length of tiling as above × two for both roof slopes |

Stop-ends to gutter

| R.10.11.2.1.0
nr. | Number of stop-ends |

Continued

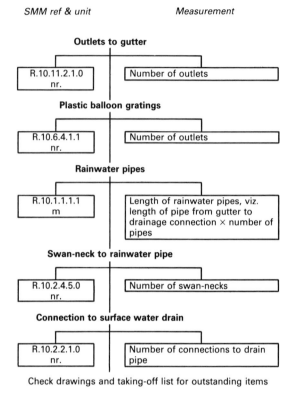

SMM ref & unit	Measurement

Outlets to gutter

| R.10.11.2.1.0
nr. | Number of outlets |

Plastic balloon gratings

| R.10.6.4.1.1
nr. | Number of outlets |

Rainwater pipes

| R.10.1.1.1.1
m | Length of rainwater pipes, viz. length of pipe from gutter to drainage connection × number of pipes |

Swan-neck to rainwater pipe

| R.10.2.4.5.0
nr. | Number of swan-necks |

Connection to surface water drain

| R.10.2.2.1.0
nr. | Number of connections to drain pipe |

Check drawings and taking-off list for outstanding items

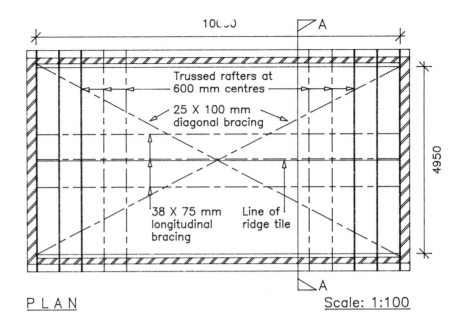

PLAN   Scale: 1:100

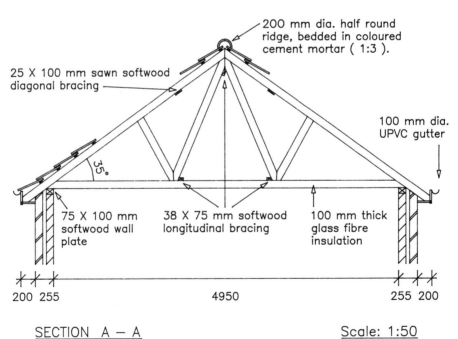

SECTION A — A   Scale: 1:50

TRUSSED RAFTER ROOF

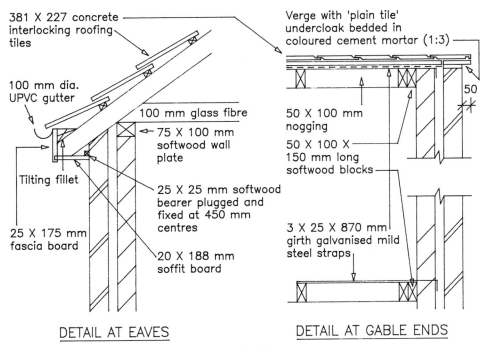

381 X 227 concrete interlocking roofing tiles

100 mm dia. UPVC gutter

Tilting fillet

25 X 175 mm fascia board

100 mm glass fibre

75 X 100 mm softwood wall plate

25 X 25 mm softwood bearer plugged and fixed at 450 mm centres

20 X 188 mm soffit board

Verge with 'plain tile' undercloak bedded in coloured cement mortar (1:3)

50

50 X 100 mm nogging

50 X 100 X 150 mm long softwood blocks

3 X 25 X 870 mm girth galvanised mild steel straps

DETAIL AT EAVES

DETAIL AT GABLE ENDS

Scale: 1:20

SPECIFICATION

1. All sawn softwood to be prime grade, treated with Tanalith C.
2. Wall plates to be fixed with 3 X 25 mm and 600 mm girth galvanised mild steel straps at 1200 mm centres one end built into brickwork.
3. Trussed rafters to be fixed to wall plate with galvanised mild steel truss clips.
4. The trussed rafters at gable ends are to be fixed to blockwork with chord straps at 2000 mm centres at rafter and ceiling joist levels as detail above.
5. The 381 X 227 mm interlocking roof tiles are to be manufactured by Messrs 'X' and laid to a 75 mm head lap, each tile nailed to 38 X 25 mm softwood battens fixed to a 306 gauge on reinforced bituminous roofing felt to BS 747 Type IF weighing 15 kg / 10 m^2.
6. All external joinery to be painted, two coats.

TRUSSED RAFTER ROOF

Drawing Number.

Name of Project.

date

Name 1

Trussed Rafter Roof 1

Taking Off List.

Construction

1. Wall plates including fixings
2. Trussed rafters   ..   ..
3. Bracing   a. horizontal
                b. raking.
4. Chord straps.
5. Insulation

Coverings

1. Tiling including battens & felt.
2. Eaves — tilting fillet.
3. Verges
4. Ridge

Eaves

1. Fascia including priming backs.
2. Soffit   ..   ..   ..
3. Ventilation
4. Boxed ends including priming backs.
5. Decoration

Rainwater Goods

1. Gutters including fittings
2. Rainwater pipes .. ..

	Page Nrs.
1	11
2	12
3	13
4	14
5	15
6	16
7	17
8	18
9	19
10	20

Name 2   date   Name of Project.

Trussed Rafter Roof 2

Construction
Plates

joint in length	10·000
	0·150
	10·150

The wall plate exceeds 6.00 m in length but it is not necessary to have it made of one length of timber. Therefore an allowance is made for a halved joint.

2/10·15   100 × 75 mm Swn. Tanalised swd. plates bedded in g.m. (1:1:6)

SMM G20.8.0.1.0

fxg. straps
1·200) 10·000
= 8 + 1 end
= 9 Nr.

2/9   Gal. m.s. wall plate fxg. straps 3 × 25 and 600 mm girth one end bent for b.i. & four times holed

SMM G20.20.1.0.0

The fixing of the strap is deemed to be included. (See SMM G20.20–C3.)

Trussed   Rafter   Roof   3

<u>Constr. (Ctd)</u>
<u>Trussed rafters</u>
<u>Nr.</u>
10·000

−gap   0·050
−¢ truss ²/0·038: 0·019
         ²/0·069: 0·138
         0·600) 9·862
         = 16+1rem+1end
         =   18 Nr.

<u>Span</u>
Int. bwk. dim.   4·950
+ wall plate ²/0·100: 0·200
                 5·150

<u>Length</u>
            4·950
+wall   0·255
+overhang 0·200
−fascia 0·025: 0·175
   ²/0·430: 0·860
            5·810

<u>Hgt</u>
= Tan 35° × 2·905
   =   2·034.

The dimension between the centres of the first and last trussed rafters is calculated, which is then divided by the centres of the trussed rafters to give the number of gaps between rafters, to which one is added to give the number of trussed rafters. If there is any remainder whatsoever an extra one is added, otherwise the distance between trussed rafters will be greater than that designed.

The span of the roof for specifying trussed rafters is measured to the external face of the wall plates.

Name of Project.   Name of

date

Name 3

Trussed Rafter Roof 4.

Constr. (Ctd)
Trussed rafters (Ctd)

18	Swn.'Tan.' swd. trussed rafters fink pattern 5·810 × 2·034 m. high %. 35° pitch, 5·150 m span with 175 mm eaves overhang both ends, 38 × 100 mm members jointed with 18g. gal. m.s. plate connectors.

SMM G20.2.1.0.0

It is good practice to include a bill diagram, and refer to it in the dimensions.

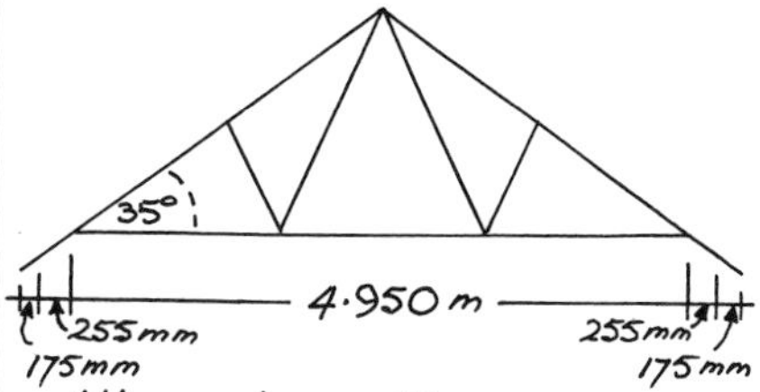

35°

— 4·950 m —

255 mm   255 mm
175 mm   175 mm

All members 38 × 100 mm treated softwood fixed with galvanised mild steel plates

BILL DIAGRAM NR. 1

Fixing rafters to wall plate

7/18	Gal. m.s. truss clips a.s. to fit 38 mm thick trussed rafters.

SMM G20.28.1.0.0

Longitudinal Bracing

	10·000
	joint in running length 0·150
	10·150

The bracing needs to act throughout its length. Therefore an allowance for a joint in the length is made. (See previous note, Timbers > 6 m.)

3/10·15	75 × 38 mm. Swn.'Tan.' swd. rf. members pitched.

SMM G20.9.2.1.0

Left margin (vertical): Name 4   date   Name of Project.

Trussed   Rafter   Roof 5.

Constr. (Ctd)

Raking bracing.

The raking bracing is indicated on the roof plan and section, but it is fixed to the underside of the rafters going from the highest midway point of the roof to the lowest part next to the gable wall.

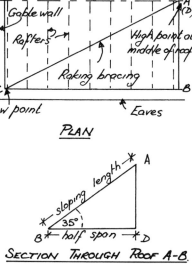

PLAN

SECTION THROUGH ROOF A-B.

sloping length A-B : $\dfrac{\text{half span}}{\text{Cos } 35°}$

$= \dfrac{\frac{1}{2}/4.950}{\text{Cos } 35°}$

$= \underline{3.021 \text{ m}}$.

length of raking brace

$= \sqrt{BC^2 + \text{sloping length}^2}$

$= \sqrt{5.000^2 + 3.021^2}$

$= \underline{5.842 \text{ m}}$

The calculation of its length is the same as for a hip rafter. (See previous example.)

SMM G20.9.2.1.0

$\dfrac{2}{2}\Big/5.84$

100 × 25 mm Swn. 'Tan.' swd. rf. members pitched.

Name of Project.

date

Name 5

Name

Trussed Rafter Roof 6.

Constr. (Ctd)
<u>Chord fixing straps</u>
<u>at gable ends</u>

<u>Ceiling jsts.</u>   <u>Number</u>
2·000) 4·950
: 2 + 1 end
= 3 Nr.

<u>Girth.</u>
end   0·075
innerskin   0·103
block   0·050
length   0·638
            0·866

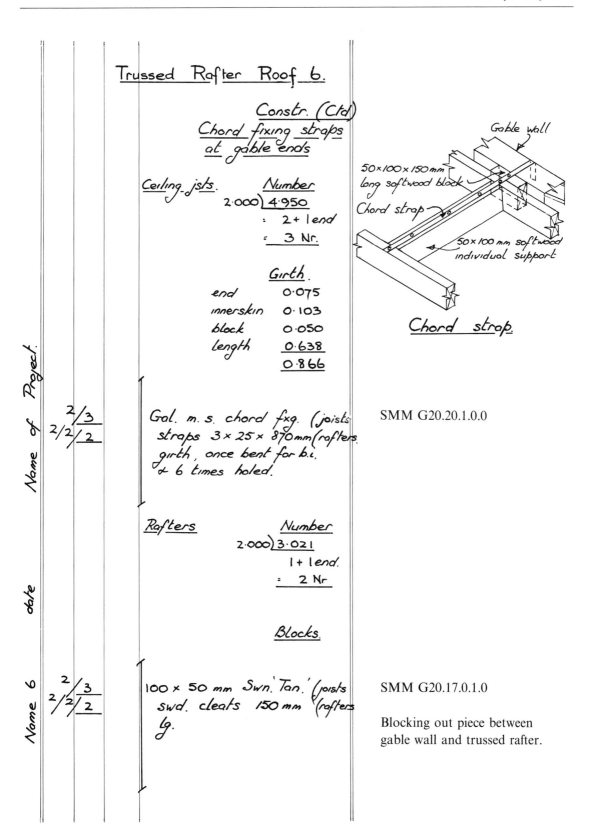

Gable wall

50×100×150mm
long softwood block

Chord strap

50×100mm softwood
individual support

<u>Chord strap.</u>

2/3
2/2/ 2

Gal. m. s. chord fxg. (joists
straps 3 × 25 × 870mm (rafters.
girth, once bent for b.i.
& 6 times holed.

SMM G20.20.1.0.0

<u>Rafters</u>   <u>Number</u>
2·000) 3·021
1 + 1 end.
= 2 Nr

<u>Blocks.</u>

2/3
2/2/ 2

100 × 50 mm Swn. Tan.' (joists
swd. cleats 150 mm (rafters
lg.

SMM G20.17.0.1.0

Blocking out piece between
gable wall and trussed rafter.

Name of Project.

date

Name 6

Trussed Rafter Roof 7.

Constr. (Ctd)
Individual supports
                                     0·600
- trussed rafters. 2/½/ ³/0·038: 0·038
                                     0·562

2/3/0·56   100 × 50 mm Swn. 'Tan ' (joists
2/2/2/0·56 swd. individual   (rafters
            supports.

Insulation

10·00   100 mm. Glass fibre
4·95   insulation quilt laid
         between members at 600
         mm. centres. horizontally,
         butt joints.

SMM G20.13.0.1.0

Support under straps between
trussed rafters.

SMM P10.2.3.1.0

Name of Project.

date

Name 7

Trussed Rafter Roof 8.

Coverings.

Length.   10·000
+ gable walls ²⁄₀·255 0·510
+ verges   ²⁄₀·050 0·100
                      10·610

Sloping length
  half span ²⁄₄·950 : 2·475
+ ext wall   : 0·255
+ overhang   : 0·200
+ ℄ gutter ½/0·100 : 0·050
      half span   2·980

slpg. length = half span
                 Cos 35°

             = 2·980
                Cos 35°

slpg. length = 3·638

HALF SECTION THROUGH ROOF.

²/10·61	Rf. cvgs 35° pitch 381 × 227mm
3·64	conc. interlocking rf. tiles

Rf. cvgs 35° pitch 381 × 227mm
conc. interlocking rf. tiles
as spec. with 75 mm head
lap, each tile in every course
nailed with one 118. 50mm.lg.
alum.alloy nail to 38 × 25 mm
swn. Tan. swd. battens fixed
to 306 mm. gauge with gal.
iron nails 50 mm lg. on a
inc. rfcd. bit. rfg. felt
to BS. 747 type IF weighing
15 kg. / 10 m² fxd. with gal.
clout hdd. nails with 150mm
laps.

SMM H60.1.1.0.0

The length times the sloping
distance of the roof covering
gives the area of one side of
the roof. This is then timesed
by two to give the area of both
roof slopes.

'As spec' (as specified)
indicates that the full
specification of the roof
tiles is included in the
Preamble Bill, and thus there
is no need to repeat it here.

Name of Project.

Name 8   date

Trussed   Rafter   Roof 9

Cvgs. (Ctd)
Eaves.

	tilting fillet.	10·610
	− verge	0·050
	− outerskin	0·103
	²⁄0·153 = 0·306
	10·304

With fully interlocking tiles there is no additional work required at the eaves and therefore no item is measured.

²⁄10·30   Swn. 'Tan' swd individual support 100 × 50mm extreme triangular shape

SMM G20.13.0.1.4

Verges

²⁄3·64   Verges on interlocking tiles including plain tile under- -cloak bedded & ptd in col. c.m. (1:3)

SMM H60.5.0.0.0

&

Left hand verges   ditto.

SMM H60.5.0.0.0

One edge of interlocking tiles is irregular and uncoloured, which would show on the verge. Special left-hand verge tiles are made to give a fair edge to the verge.

Name of Project.

date

Name 9

Trussed Rafter Roof 10.

Cvgs. (Ctd)
Ridge.

| | 10.61 | 200 mm Dia. h.r. ridge tiles to match general tiling bedded & ptd. in col. c.m. (1:3) | SMM H60.6.0.0.0 |

Eaves
10.610
- verge overhang 2/0.050 0.100
10.510

| 2/ | 10.51 | 25 × 175 mm Wrot swd. fascia bds. width ≤ 300 mm, once grooved. | SMM G20.15.3.2.0 |

& overhang 0.200
- fascia 2/0.025 0.013
0.187

| | | 20 × 187 mm Ditto eaves or verge soffit bds. width ≤ 300 mm, once rebated. | SMM G20.16.3.2.0 |

&

| | | 25 × 25 mm Swn. Tan. swd. individual support p.& fxd. to bwk. at 450mm centres. | SMM G20.13.0.1.0 |

Left margin (vertical): Name of Project. | Name | date | Name 10

Trussed   Rafter   Roof 11

<u>Eaves</u> *(Ctd)*

2/2/ 10·51

Prime only gen.   (Fascia
surf. of swd.   (x2 for
isol. surf. girth   \ soffit
≤ 300 mm. application
on site prior to fxg.

SMM M60.1.0.2.4

All external joinery should be
protected from the weather
by painting or priming the
backs before fixing.

<u>To Take</u>
eaves   ventilation system.

2/2/1

25 mm Wrot swd. spandril
shaped boxed end to
eaves   200 × 275 mm
overall.

SMM G20.18.0.1.0

&

Prime only gen. surf. of
swd. isol. areas ≤ 0·50 m²
irrespective of girth
application on site
prior to fxg.

SMM M60.1.0.3.4

Name 11   date   Name of   Name of Project.

<u>Trussed Rafter Roof 12</u>.

<u>Eaves (Ctd)</u>
<u>Decoration.</u>

fascia	0.175
overhang	0.200
recess	0.025
Girth of eaves.	0.400

The fascia and soffit boarding can be painted at the same time and therefore these are girthed together.

2/ 10.51
2/3/ 0.40
0.20
0.28

K.P.S & ② on gen. surf. of swd. girth > 300 mm (Ends

<u>Ext.</u>

SMM M60.1.0.1.0

Painting is deemed to be internal. (See M60–D1.) Therefore external work must be described as such.

<u>Rainwater Goods</u>

2/ 10.61

U.P.V.C. rainwater gutters manu. by Messrs 'B' str. 100mm dia. h.r. with union clip jts. fxd. with bkts. at 1m centres to swd. with 8g. 50mm lg. gal. round headed screws.

SMM R10.10.1.1.1

Gutters are measured over all fittings (SMM R10–M6) and include joints in the running length (SMM R10–C9).

2/2/ 1

Extra over ditto for stop ends.

SMM R10.11.2.1.0

2/ 1

E. o. ditto for outlet with nozzle for 75mm dia. pipe.

SMM R10.11.2.1.0

Name of Project.   *Name 12   date   Name of Project.*

Trussed Rafter Roof 13.

Rainwater goods (Ctd)

2/1	U.P.V.C. balloon grtg. & fxg. in 75mm dia outlet.	SMM R10.6.4.1.1

Pipes

| 2/6.00 | U.P.V.C. rainwater (length pipes manu. by assumed) Messrs 'B', str. 75mm dia. with socketted push fit jts. fxd with straps at 2m. centres to faced bwk. with 89, 75mm lg. gal. m.s. round headed screws. | SMM R10.1.1.1.1 |

Pipes are measured over all fittings (SMM R10–M1) and include joints in the running length (SMM R10–C3).

| 2/1 | E.o. ditto for swanneck 200mm projection | SMM R10.2.4.5.0 |

&

| | E.o. ditto for special jt. & connection to back inlet of earthenware gulley including caulking bush & cmt. jt. | SMM R10.2.2.1.0 |

Name of Project

date

Name 13

Name

11

Built-up felt flat roof

Generally

Roofs with a pitch of 10 degrees or less are termed flat roofs. With regard to the measurement of the structure, this chapter should be read in conjunction with the comments made in Chapter 9.

Measurement

Strutting

If the span of the roof joists exceeds 2.4 m then a row of strutting should be measured in linear metres between wall faces and over the joists, see SMM G20.10.1 or 2.1. The strutting helps to prevent the joists twisting or flexing in their length.

Firrings

Firrings are pieces of timber cut raking one edge under building or decking, either fixed to the top edge of each joist or at right angles to them depending on the required direction of slope. Firrings give a flat roof sufficient fall to enable rainwater to be discharged and are measured in linear metres and described as individual supports with their width and average depth stated, see SMM G20.13.

Coverings

As the name implies, the roof covering is built up from several layers of felt bonded together and to the base, which is generally finished with reflective paint, sand, stones or special slabs, depending on the type of use and amount of access. The felt is measured the area in contact with the base in square metres, see SMM J41.1–M2.

   Perimeter work is measured linear if the girth of the work does not exceed 2 m or superficially if the girth of the work exceeds 2 m. Preformed metal edge trim is measured in accordance with SMM J41.19 and any lead cover flashings are measured in linear metres in accordance with SMM H71.10.

Flow chart for built-up felt flat roof

SMM ref & unit		*Measurement*
	Start	

Roof joists

G.20.9.1.1.0 m	Length of roof joists, viz. length of each roof joist × number of roof joists at 300 mm centres

Joist hangers

G.20.21.1.0.0 nr.	Number of joist hangers, viz. number of joists as above

Herringbone strutting

G.20.10.1.1.0 m	Length of strutting measured over joists, viz. length measured to external faces of first and last joists

Firring pieces

G.20.13.0.1.5 m	Length of firring pieces, viz. length measured from brickwall to outside face of batten attached to fascia (fixed on top of roof joists) × number of roof joists as used above

Roof decking

G.32.1.1.0.0 m²	Area of woodwool decking, viz. length measured to external face of first and last joists × width (same as length of firring piece)

Kerbs

G.20.13.0.1.5 m	Length of kerbs, viz. length of kerb × two for both ends of roof

Angle fillet

G.20.13.0.1.4 m	Length of angle fillet, viz. length next brickwall measured between inside face of kerbs plus length next kerb × two for both ends plus length to end of kerb × two for both ends

Insulation

To take note	Not measured in this example

Fascia boards at front

G.20.15.3.2.0 m	Length of built up fascia board behind gutter, viz. length measured between external face of fascias at sides

Continued

SMM ref & unit		*Measurement*

Fascia boards at sides

G.20.15.2.2.0 m	Length of built up fascia boards at sides, viz. length measured between brickwork and external face of fascia at front × two for both ends of roof

Prime only back of fascia boards

M.60.1.0.2.4 m	Length of boards making up all fascias, viz. length of front fascia × number of boards making up fascia plus length of side fascia × number of boards making up side fascia × two for both ends of roof

Painting fascia boarding not exceeding 300 mm girth

M.60.1.0.2.3 m	Length of external irregular general surfaces of front fascia, viz. length of front fascia

Ditto exceeding 300 mm girth

M.60.1.0.1.3 m²	Area of external irregular general surfaces of side fascias, viz. length of side fascia × girth of exposed face of side fascia × two for both ends of roof

Roof covering

J.41.2.1.0.0 m²	Area of felt measured in contact with base, viz. length of roof measured between inside edges of angle fillets at kerbs × width of roof measured from inside edge of angle fillet at brickwall to outside face of bearer fixed on fascia

Skirting

J.41.10.2.0.0 m	Length of skirting next brickwall, viz. length measured between inside face of kerbs

Lead cover flashing

H.71.10.1.0.1 m	Length of lead cover flashing next to brickwall, viz. overall length of roof, plus 150 mm overlap at each end

Continued

Flow chart for built-up felt flat roof – continued

SMM ref & unit   Measurement

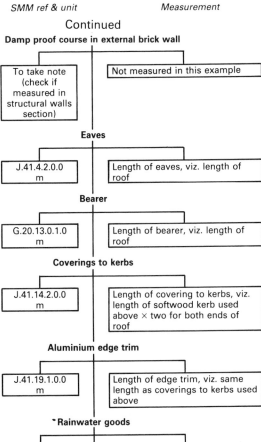

Continued

Damp proof course in external brick wall

To take note (check if measured in structural walls section) — Not measured in this example

Eaves

J.41.4.2.0.0 m — Length of eaves, viz. length of roof

Bearer

G.20.13.0.1.0 m — Length of bearer, viz. length of roof

Coverings to kerbs

J.41.14.2.0.0 m — Length of covering to kerbs, viz. length of softwood kerb used above × two for both ends of roof

Aluminium edge trim

J.41.19.1.0.0 m — Length of edge trim, viz. same length as coverings to kerbs used above

Rainwater goods

To take note — Not measured in this example

Check drawings and taking-off list for outstanding items

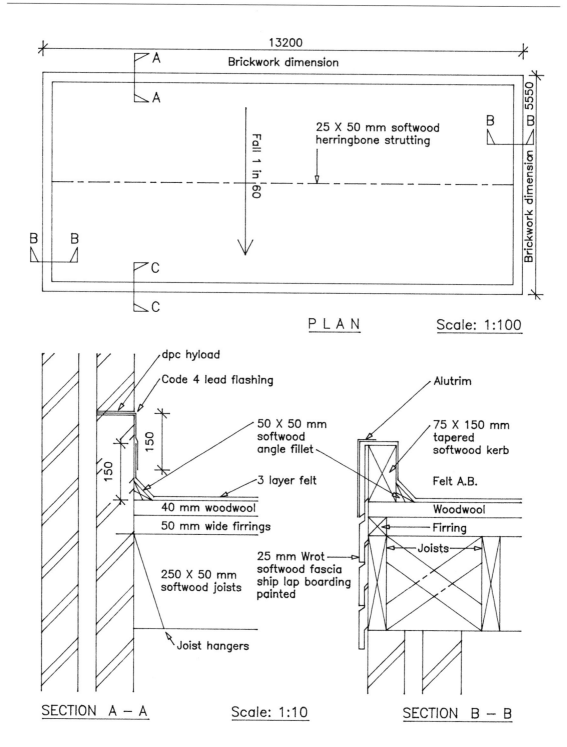

PLAN　　　Scale: 1:100

13200
Brickwork dimension

A

A

25 X 50 mm softwood
herringbone strutting

B　B

5550
Brickwork dimension

Fall 1 in 60

B　B

C

C

SECTION A — A　　　Scale: 1:10　　　SECTION B — B

dpc hyload

Code 4 lead flashing

150

150

150

50 X 50 mm
softwood
angle fillet

3 layer felt

40 mm woodwool

50 mm wide firrings

250 X 50 mm
softwood joists

Joist hangers

25 mm Wrot
softwood fascia
ship lap boarding
painted

Alutrim

75 X 150 mm
tapered
softwood kerb

Felt A.B.

Woodwool

Firring

Joists

FLAT ROOF

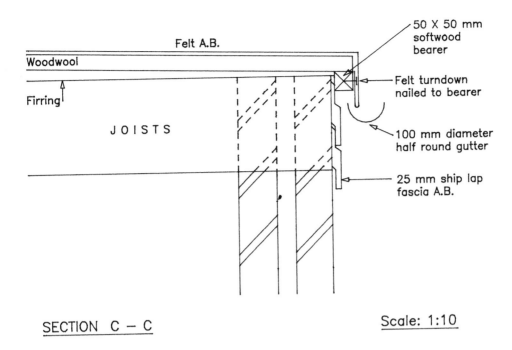

Felt A.B.

Woodwool

Firring

J O I S T S

50 X 50 mm
softwood
bearer

Felt turndown
nailed to bearer

100 mm diameter
half round gutter

25 mm ship lap
fascia A.B.

SECTION C – C

Scale: 1:10

SPECIFICATION

1. Three layer built up felt roofing.
   1st layer 'Rubervent' glass fibre weighing 31.7 kg / 10 m²
   placed on woodwool.
   2nd and 3rd layers 'Rubervent glasphalt' weighing 18.1 kg /
   10 m² hot bitumen bonding between layers.
   Finish with white spar chippings bedded in hot bitumen.
2. 250 X 50 mm stress graded softwood tanalised joists at
   300 mm centres with one row of 50 X 25 mm herringbone
   strutting.
3. 40 mm 'Woodcemair' unreinforced prescreeded top surface
   fixing galvanised roof slab nails at 150 mm centres.
4. 'Alutrim' extruded aluminium roof trim type X to receive built
   up felt roofing with sleeved joints and screwed to softwood
   at 450 mm centres.
5. Lead flashing is code 4 to BS.1178 wedged into groove in
   brickwork with lead wedges. Lapped 150 mm at laps.

F L A T   R O O F

Built Up Felt Roof 1

Taking Off List.

Construction

1. Joists
2. " hangers
3. Strutting
4. Firrings
5. Decking
6. Kerbs
7. Angle fillet
8. Insulation

Eaves

1. Fascias including priming.
2. Decoration

Coverings.

1. Felt roofing.
2. Skirting
3. Flashing
4. Eaves
5. Bearer
6. Covering to kerbs.
7. Edge Trims.

Rainwater Goods

1. Gutter including fittings
2. Pipes " "

Drawing Number.

Name of Project

date

Name 1

Page	Nrs
1	11
2	12
3	13
4	14
5	15
6	16
7	17
8	18
9	19
10	20.

Name of Project.

Name 2 date

<u>Built Up Felt Roof 2</u>

<u>Construction</u>

N.B.
Check that the brick cavity wall above roof has been measured with structural walls

Note to check that the wall has already been measured and is not to be measured in the roof section.

<u>Joists</u>

	<u>Number</u>
	13·200
2/½/0.050:	0·050
0·300)	13·150
	43+ lrem.+ lend.
	= 45 Nr.

-¢/joists

To calculate the number of joists, find the dimension between the centre lines of the first and last joists, divide by the centres of the joists, add one for any remainder and add one for the end to give total number. If the length of the joists exceeds 6 m and they are required to be in one continuous length then this must be stated in the description. (See SMM G20.6.0.1.1.)

45/5·55

50 × 250 mm Swn. 'Tanalised' stress graded swd. as spec. roof members flat.

SMM G20.9.1.1.0

45

Gal. m.s. jst. hangers manufactured by MacAndrews & Forbes to suit 50×250mm. jsts.

SMM G20.21.1.0.0

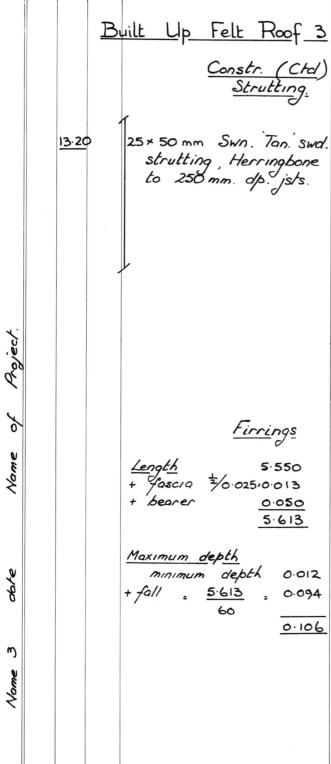

Built Up Felt Roof 3

Constr. (Ctd)
Strutting:

13.20 | 25 × 50 mm Swn. Tan.' swd. strutting, Herringbone to 250 mm. dp. jsts.

SMM G20.10.1.1.0

Strutting is measured over the joists. (See SMM G20–M1). The one measurement includes for both rows in the case of herringbone strutting.

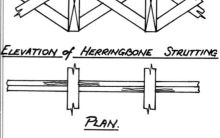

ELEVATION of HERRINGBONE STRUTTING

PLAN.

Firrings

Length	5.550
+ fascia ½/0.025·0.013	
+ bearer	0.050
	5.613

Maximum depth	
minimum depth	0.012
+ fall = 5.613/60 =	0.094
	0.106

Length of firrings which are fixed on top of and supported by the joists is calculated. Once the length of firring is known the maximum depth can be calculated by using the proportions of the fall of the roof (in this case 1 in 60).

ELEVATION OF FIRRING.

The firring will always have a minimum depth for fixing

Name A   date   Name of Project.

Built Up Felt Roof 4

Constr. (Ctd)
Firrings (Ctd)

purposes of 12 mm at the end
next to the gutter. Then
dividing the length of the
firring by 60 will
give a dimension which is
equal to the required fall.
This dimension is then added
to the minimum depth to give
the depth of timber required
to cut the fixing piece.

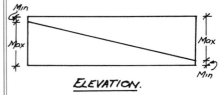

ELEVATION.

In practice one piece of
timber of a depth slightly in
excess of the maximum will be
cut to provide two fixings.
This type of firring is
described as tapered and
runs with the fall of the roof.
Where the joists run at right
angles to the fall of the roof,
the firrings are described as
splayed and each one will be
a different height, although for
measurement purposes they are
measured together and the
average depth is stated.

SMM G20.13.0.1.5

The number is not required by
SMM but is given to help the
estimator price the cutting.

$\dfrac{45}{5\cdot 61}$

Swn. 'Tan.' swd. individual
supports 50 mm. wide 106mm.
max. depth tapering to
12 mm min. depth (In. 45 Nr)

<u>Built Up Felt Roof 5</u>

<u>Constr (Ctd)</u>
<u>Decking:</u>

13·20 5·61	Edge supported woodwool slab decking 40mm thick manu. by 'Woodcemair' prescreeded fin. top fxd. with gal. rf. slab nails a.s. to swd. jsts. at 300mm centres, butt jts.

SMM G32.1.1.0.0

<u>Kerbs</u>

Max. depth of kerb	0·150
– fall in roof	0·094
Min. depth.	0·056

The kerbs are set on the decking which falls towards the gutter, therefore in order to have a level top to the kerb they are tapered the same amount as the firrings but in the opposite direction.

2/5·55	Swn. 'Tan.' swd. individual supports 75 mm wide, 150mm. max. depth tapering to 56 mm. min. depth. (In 2 Nr.)

SMM G20.13.0.1.5

Name of Project.

Name 6   date   Name of Project.

Built Up Felt Roof 6

Constr (Ctd)
Angle fillet.

Girth.   next bwk.   13·200
            – kerbs 2/0·075: 0·150
                               13·050
+ next kerbs   2/5·550: 11·100
+ end of kerbs   2/0·075: 0·150
+ passings 2/2/2/½/0·050: 0·200
                              24·500

24·50 | 50 × 50 mm Extreme
            triangular swn.'Tan.' swd.
            individual supports.

To Take
Insulation.

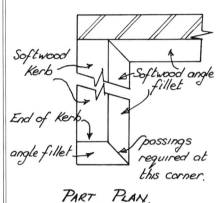

Softwood
Kerb   Softwood angle
                  fillet

End of Kerb.
angle fillet   passings
                      required at
                      this corner.

PART   PLAN.

SMM G20.13.0.1.4

Insulation is not shown on the
drawing, but is required to
comply with the Building
Regulations.
Therefore a 'To Take' note is
added to the dimension so that
it will not be forgotten, and
an item is put into the query
book.

Name 7   date   Name of Project.

Built Up Felt Roof 7.

Eaves

Sides   Length
                          5·550
+ ext. angle   0·025
                          5·575

                              Height.
Behind gutter. joist : 0·250
+ overlap of bwk. say: 0·050
                          0·300

Sides   front a.b.   0·300
             firring min.   0·012
             decking   0·040
             kerb max.   0·150
                             0·502

Front   Length
                          13·200
+ ext. angles ²⁄0·025: 0·050
                          13·250

	13·25	25mm Wrot swd. (front fascia boards width ≤ 300 mm, ship lap bdg. 300 mm. high comprising 200 mm. wide bds. secret nailed.

SMM G20.15.3.2.0

	2/5·58	25mm Ditto width (sides > 300mm, ditto, 502mm high ditto.

SMM G20.15.2.2.0

The external angle allowance is added to the length for an external mitred angle with the front fascia.

Name 8 date Name of Project.

Built Up Felt Roof 8.

Eaves (Ctd)

Priming backs of bds.

NB. Cover of one bd. say 170mm.

front fascia $\frac{0.300}{0.170}$ = 2 bds.

side fascia $\frac{0.502}{0.170}$ = 3 bds.

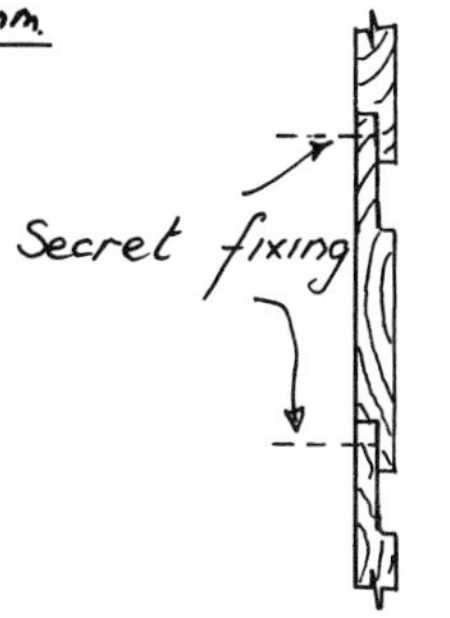

SECTION THROUGH SHIP LAP BOARDING.

The fascias are built up of a series of boards and the back and edges of each board need to be primed before it is fixed. Therefore the number of boards in each fascia must be calculated.

2/	13.25	Prime only gen. surf. (front of swd. Isol. surf. (sides girth ≤ 300 mm application on site prior to fixing.
2/3/	5.58	

SMM M60.1.0.2.4

Decoration

Front

hgt.	0.300
edges 2/2/0.013 =	0.052
	0.352
− batten at top:	0.050
	0.302

The area measured for painting must include an allowance for the extra girth of edges etc. (See SMM M60–M2.) The exposed edges and an equivalent allowance for the moulding is added for each board to give the girths of the fascias.

Name of Project.

date

Name 9

Built Up Felt Roof 9

Eaves (Ctd)
Decoration (Ctd)

Side   hgt.   0·502
+ edges ³/²/0·013: 0·078
0·580
- trim   0·050
0·530

	13·25	K.p.s & paint 2 under-(front -coats and 1 finishing ct. of paint on swd. Isol. surf. girth ≤ 300mm irregular surf. *external*

SMM M60.1.0.2.3

The painting of the fascias has to be described as irregular surfaces. (See SMM M60–D4.)

2/	5·58 0·53	Ditto on swd. gen. (side surf. girth > 300mm irreg. surf. *external*

SMM M60.1.0.1.3

Coverings

13·200
- kerb   0·075
- L. fillet   0·050
2/ 0·125   - 0·250
12·950

The area measured is that in contact with the base. (See SMM J41–M2.)

<u>Built Up Felt Roof 10.</u>

<u>Cvgs. (Ctd)</u>

		5·550
+ fascia	½/0·025:	0·013
+ bearer	:	0·050
		5·613
- ∠. fillet	:	0·050
		5·563

12·95	Three layer built up felt
5·56	flat rf. cvgs. 1st layer

'Rubervent' glass fibre
felt weighing 31·7 kg/
10m² placed on prescreeded
woodwool decking 2nd
& 3rd. layers 'Rubervent'
glass fibre felt weighing
18·1 kg/10m² hot bit.
bonded between layers
finished with white spar
chippings bedded in
hot. bit.

SMM J41.2.1.0.0

<u>Skirting</u>

<u>Girth</u>

height		0·150
- ∠. fillet		0·050
		0·100
+∠ fillet : √0·050²+0·050²:	0·071
		0·171

Name 10   date   Name of Project.

Built Up Felt Roof 11

Cngs. (Ctd)
Sktgs.(Ctd)

13·20	3 Layer built up felt o.b. venting sktg. with top layer in green mineralised felt sktg. girth ≤ 2·00m & n. e. 200 mm. dressed over ∠ fillet at bttm.	SMM J41.10.2.0.0

The skirting is notched over the top of the kerbs at each end and is deemed to be included. (See SMM J41–C1(b).)

Lead flashing.

Length		13·200
+ returns at ends²/0·150:		0·300
		13·500

The lead flashing is taken 150 mm beyond the ends of the felt skirting.

Girth	hgt.	0·150
+ turn in		0·025
		0·175

13·50	Lead sheet flashing Code 4 to B.S. 1178. 175 mm girth horiz. Lead wedged into jt. of bwk. inc. Code 5 50 mm wide lead tacks at 450 mm centres, lapped 150 mm at all passings.	SMM H71.10.1.0.1

Name 11   date   Name of Project.

Built Up Felt Roof 12.

Cvgs. (Ctd)

To Take
Dpc. in ext. wall - check
if measured with wall.

	Eaves
Girth	0·150
turn under	0·100
	0·250

13·20 — 3 Layer built up felt a.b.
with top layer green
mineralised, eaves, girth
≤ 2·00m, exc. 200 &
n.e. 400mm gal. clout
nailed to swd. at 100mm
centres to form drip.

&

50 × 50 mm Swn. Tan. swd.
individual support.

SMM J41.4.2.0.0

SMM G20.13.0.1.0

Name of Project.

Name of Project.

date

Name 12

Name

<u>Built Up Felt Roof 13</u>

<u>Cvgs (Ctd)</u>
<u>Kerbs.</u>

<u>Girth</u>	top		0·075
	+ into trim		0·025
side	max	0·150	
	min	<u>0·056</u>	
	2)	0·206	
	av ∶	0·103	
− ∠. fillet		<u>0·050</u>	
		0·053	
+ ∠. fillet a.b	0·071	:	<u>0·124</u>
			0·224

2/	<u>5·55</u>	3 Layer built up felt a.b. with top layer green mineralised felt cvgs. to kerbs girth ≤ 2·66m, exc. 200 & n.e. 400 mm. dressed into aluminium flashing & over angle fillet at bttm.	SMM J41.14.2.0.0

Covering to the ends of the kerb is deemed to be included. (See SMM J41–C2.)

&

Preformed extruded aluminium edge trim manu. by 'Alutrim' Type 'X' to receive b.u. felt with sleeved jts. Screwed to swd. at 450 mm. centres with countersunk aluminium screws	SMM J41.19.1.0.0

<u>To Take</u>
All rainwater goods.

(left margin, vertical:) Name of Project.　date　Name 13　Name

12

Reinforced concrete flat roof

Introduction

The notes made in previous chapters where applicable also apply to this example, with the following additional comments.

Measurement

In-situ *concrete*

Concrete in slabs is measured in cubic metres with the thickness range stated in the description. If the concrete is reinforced then this must be stated, see SMM E10.5.1–3.0.1.

Any beam or beam casings attached to the soffit which would normally be poured at the same time as the slab are measured in cubic metres and are included with the slab, see SMM E10.5–D4, although the thickness range of the slab over the beam should not be adjusted. Attached beams or beam casings are only measured separately if they are defined as deep beams, see SMM E10.9–D7 and Figures 12.1 and 12.2.

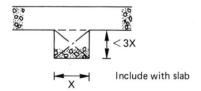

Figure 12.1   *Section*

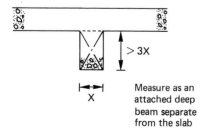

Figure 12.2   *Section*

Formwork

Formwork is measured to the concrete surfaces of the finished structure which require temporary support during casting, see SMM E20–M1. Formwork to the soffit slabs is measured in square metres and the slab thickness and height to soffit are given in the description, see SMM E20.8.1.1.1, etc. Formwork to regular shaped (i.e. a rectangle or square in cross-section) beams is measured in square metres and the number of beams and height to soffit are given in the description, see SMM E20.13.1.1.1. Formwork to edges of suspended slabs can include formwork to the side of upstands if they are in the same plane and is measured in linear metres if less than 1 m high. Formwork also provides the finish to the concrete surface and, therefore, if any finish other than the basic finish is required, it is measured 'extra over the basic finish', see SMM E20.20, etc.

Reinforcement

The design engineer usually produces a schedule for all reinforcement which will give the bar type, diameter, number of bars, shape of bars, dimension of bars and length of bar needed to form the required shape, i.e. to make allowance for hooks, bends, etc. This schedule can be used to weight up the bars for direct entry in the dimensions or abstracted for billing in tonnes in the bills of quantities. However, in the absence of such a schedule bar reinforcement is measured in linear metres, with its diameter and shape stated and then weighted up.

In the following example, the number and lengths of bar reinforcement have been calculated from the drawing by first deducting the amount of concrete cover (for weather or fire protection) from both sides and then adding an allowance of nine times the diameter of the bar for each hook. Often, to increase the bond between steel and concrete, twisted or formed bars are used. BS4466 lists the various hook shape references. The length of the bar before bending must also be considered as bars exceeding 12 m in length and fixed horizontally, or bars exceeding 6 m long which are fixed vertically, must be given in 3 m stages, see SMM E30.1.1.1–4.

Fabric reinforcement is measured in square metres stating the mesh size and weight per square metre. All reinforcement is deemed to include tying wire, spacers and chairs which hold the reinforcement the required distance above formwork or apart from each other, see SMM E30.1 or 4 and C1 or C2 respectively.

Screed

In the following worked example the insulation and falls to the roof are provided by the screed which directs the rainwater to outlets. Consequently, the falls in the surface of the screed are in two directions and are described as to falls and cross falls. The dimensions measured are those in contact with the base and given in square metres if exceeding 300 mm wide with the average thickness and number of coats being given in the description, see SMM M10.5.

Asphalt roofing

Asphalt roofing is measured in square metres, being the area in contact with the base. The description will include the width in stages, see SMM J20.3.1–4, with the thickness, number of coats, any surface treatment and the nature of the base to which applied stated, see SMM J20–S1–S4. Coverings to kerbs, etc., are measured in linear metres and the girth is measured on face, see SMM J20.11–M4, and is deemed to include edges, arrises (angles), internal angle fillets, etc., see SMM J20.11–C5.

Flow chart for reinforced concrete flat roof

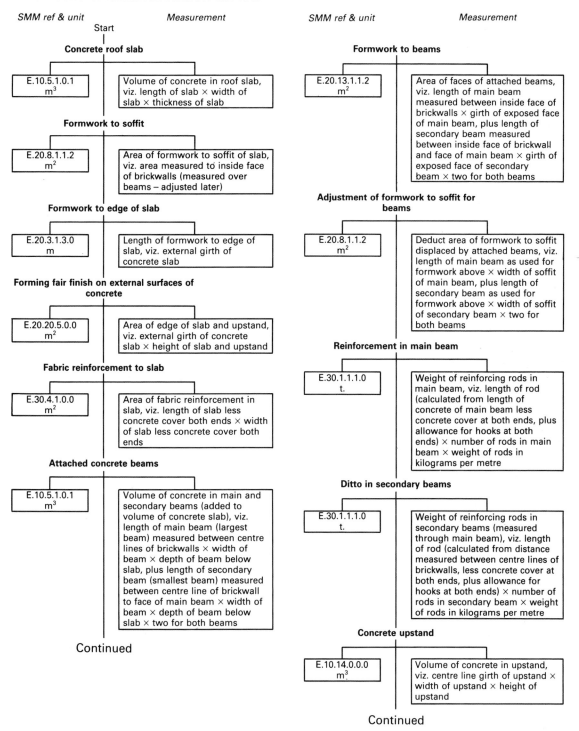

SMM ref & unit	Measurement
Start	

Concrete roof slab

E.10.5.1.0.1
m³ — Volume of concrete in roof slab, viz. length of slab × width of slab × thickness of slab

Formwork to soffit

E.20.8.1.1.2
m² — Area of formwork to soffit of slab, viz. area measured to inside face of brickwalls (measured over beams – adjusted later)

Formwork to edge of slab

E.20.3.1.3.0
m — Length of formwork to edge of slab, viz. external girth of concrete slab

Forming fair finish on external surfaces of concrete

E.20.20.5.0.0
m² — Area of edge of slab and upstand, viz. external girth of concrete slab × height of slab and upstand

Fabric reinforcement to slab

E.30.4.1.0.0
m² — Area of fabric reinforcement in slab, viz. length of slab less concrete cover both ends × width of slab less concrete cover both ends

Attached concrete beams

E.10.5.1.0.1
m³ — Volume of concrete in main and secondary beams (added to volume of concrete slab), viz. length of main beam (largest beam) measured between centre lines of brickwalls × width of beam × depth of beam below slab, plus length of secondary beam (smallest beam) measured between centre line of brickwall to face of main beam × width of beam × depth of beam below slab × two for both beams

Continued

SMM ref & unit	Measurement

Formwork to beams

E.20.13.1.1.2
m² — Area of faces of attached beams, viz. length of main beam measured between inside face of brickwalls × girth of exposed face of main beam, plus length of secondary beam measured between inside face of brickwall and face of main beam × girth of exposed face of secondary beam × two for both beams

Adjustment of formwork to soffit for beams

E.20.8.1.1.2
m² — Deduct area of formwork to soffit displaced by attached beams, viz. length of main beam as used for formwork above × width of soffit of main beam, plus length of secondary beam as used for formwork above × width of soffit of secondary beam × two for both beams

Reinforcement in main beam

E.30.1.1.1.0
t. — Weight of reinforcing rods in main beam, viz. length of rod (calculated from length of concrete of main beam less concrete cover at both ends, plus allowance for hooks at both ends) × number of rods in main beam × weight of rods in kilograms per metre

Ditto in secondary beams

E.30.1.1.1.0
t. — Weight of reinforcing rods in secondary beams (measured through main beam), viz. length of rod (calculated from distance measured between centre lines of brickwalls, less concrete cover at both ends, plus allowance for hooks at both ends) × number of rods in secondary beam × weight of rods in kilograms per metre

Concrete upstand

E.10.14.0.0.0
m³ — Volume of concrete in upstand, viz. centre line girth of upstand × width of upstand × height of upstand

Continued

Flow chart for reinforced concrete flat roof – continued

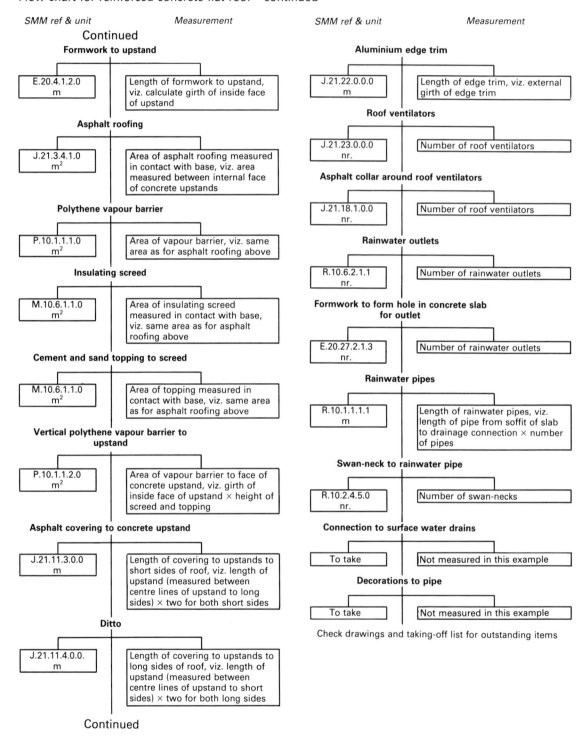

SMM ref & unit · Measurement · SMM ref & unit · Measurement

Continued

Formwork to upstand

E.20.4.1.2.0
m — Length of formwork to upstand, viz. calculate girth of inside face of upstand

Asphalt roofing

J.21.3.4.1.0
m² — Area of asphalt roofing measured in contact with base, viz. area measured between internal face of concrete upstands

Polythene vapour barrier

P.10.1.1.1.0
m² — Area of vapour barrier, viz. same area as for asphalt roofing above

Insulating screed

M.10.6.1.1.0
m² — Area of insulating screed measured in contact with base, viz. same area as for asphalt roofing above

Cement and sand topping to screed

M.10.6.1.1.0
m² — Area of topping measured in contact with base, viz. same area as for asphalt roofing above

Vertical polythene vapour barrier to upstand

P.10.1.1.2.0
m² — Area of vapour barrier to face of concrete upstand, viz. girth of inside face of upstand × height of screed and topping

Asphalt covering to concrete upstand

J.21.11.3.0.0
m — Length of covering to upstands to short sides of roof, viz. length of upstand (measured between centre lines of upstand to long sides) × two for both short sides

Ditto

J.21.11.4.0.0.
m — Length of covering to upstands to long sides of roof, viz. length of upstand (measured between centre lines of upstand to short sides) × two for both long sides

Continued

Aluminium edge trim

J.21.22.0.0.0
m — Length of edge trim, viz. external girth of edge trim

Roof ventilators

J.21.23.0.0.0
nr. — Number of roof ventilators

Asphalt collar around roof ventilators

J.21.18.1.0.0
nr. — Number of roof ventilators

Rainwater outlets

R.10.6.2.1.1
nr. — Number of rainwater outlets

Formwork to form hole in concrete slab for outlet

E.20.27.2.1.3
nr. — Number of rainwater outlets

Rainwater pipes

R.10.1.1.1.1
m — Length of rainwater pipes, viz. length of pipe from soffit of slab to drainage connection × number of pipes

Swan-neck to rainwater pipe

R.10.2.4.5.0
nr. — Number of swan-necks

Connection to surface water drains

To take — Not measured in this example

Decorations to pipe

To take — Not measured in this example

Check drawings and taking-off list for outstanding items

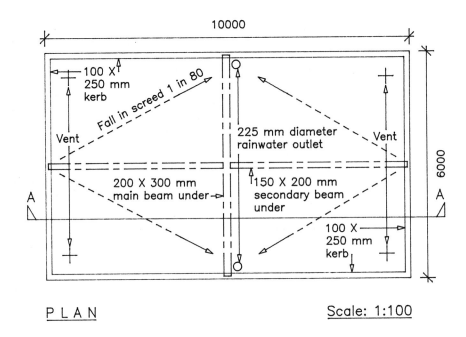

P L A N   Scale: 1:100

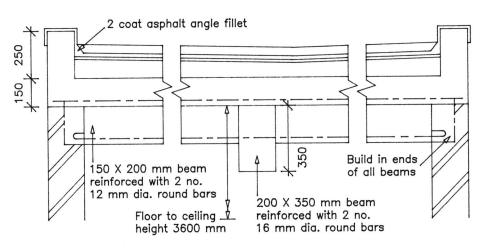

SECTION   A — A   Scale: 1:20

CONCRETE FLAT ROOF

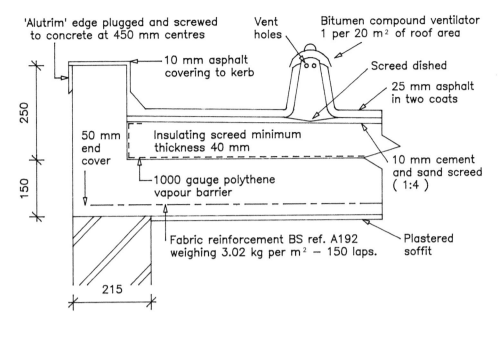

'Alutrim' edge plugged and screwed to concrete at 450 mm centres

Vent holes

Bitumen compound ventilator 1 per 20 m² of roof area

10 mm asphalt covering to kerb

Screed dished

25 mm asphalt in two coats

250

150

50 mm end cover

Insulating screed minimum thickness 40 mm

1000 gauge polythene vapour barrier

10 mm cement and sand screed (1:4)

Fabric reinforcement BS ref. A192 weighing 3.02 kg per m² — 150 laps.

Plastered soffit

215

DETAIL OF KERB AND VENT   Scale: 1:10

SPECIFICATION

1. Concrete to be 25N / 20 mm

2. External exposed concrete to have a fair face.

3. Mastic Asphalt to BS 988 limestone aggregate in two coats. Laid on sheathing felt to BS 747 4A (i), finished with white spar chippings in bitumen compound.

4. Polythene vapour barrier to have solvent welded joints.

5. The 'Alutrim' edging to have sleeved joints.

CONCRETE FLAT ROOF

Reinforced Concrete Flat Roof 1

Taking Off List

Construction

1. Slab   a.   Concrete
         b.   Formwork
         c.   Reinforcement
2. Beams   a.   Concrete
         b.   Formwork
         c.   Reinforcement
         d.   Adjustment of Formwork to slab.
3. Upstands. a   Concrete
         b.   Formwork.

Coverings

1. Asphalt to roof
2. Screeds
3. Asphalt covering to kerbs.
4. Edging.
5. Vents.

Rainwater Goods.

1. Pipe including fittings
2. Builders work.

Drawing Number.

Name of Project

date

Name 1

Page Nrs.

1	11
2	12
3	13
4	14
5	15
6	16
7	17
8	18
9	19
10	20

R.C. Flat Rf. 2.

Construction
Slab.

N.B. The external walls will
be built up to the underside
of the concrete slab with
pockets left for ends of
the beams.

10·00	
6·00	
0·15	*In situ conc. (25N/20mm*
5·79	*agg.) slabs thickness*
0·20	*≤ 150mm , rfcd.*
0·35	*(main beam*
9·59	
0·15	
0·20	*(sec. beam*

SMM E10.5.1.0.1

Formwork.

	10·000	×	6·000
−walls ³⁄₆·215	0·430		0·430
	9·570		5·570

The formwork is measured to
the inside face of the brick
external walls, over the beams,
which are adjusted later.

9·57	*Fmwk. for in situ. conc.*
5·57	*soff. of slabs, slab*
	thickness ≤ 200mm
	horiz., hgt. to soff. exc
	3·00m & n.e. 4·50m.

SMM E20.8.1.1.2

R.C. Flat Rf. 3.

Constr. (Ctd)
Slab (Ctd)
Fmwk. (Ctd)

Girth		Edge	
		10·000	
		6·000	
	2/	16·000	
	:	32·000	
Height	slab	0·150	
	upstand·	0·250	
		0·400	

As the edge of concrete slab and the outside face of upstand are in the same plane, it is sensible to measure the formwork as one item.

32·00	Fmwk. for in situ. conc. edges of suspd. slabs plain vert. hgt 250·500mm.

SMM E20.3.1.3.0

&

E.o. a basic fin. for a lining to fmwk. to produce a fair finish to edges of suspd. slabs & upstand. Sup. × 0·40 =   m²

SMM E20.20.5.0.0

400 mm is the combined height of the edge of concrete slab and upstand.

To Take
Fmwk. to int. face of Upstand.
Meas on Rf 8.

As formwork has been measured to one face of the upstand out of sequence, a note is inserted to make sure that formwork to the internal face is not forgotten.

Name of Project.   date   Name 3   Name

Name of Project. Name & date

R.C. Flat Rf. 4.

Constr. (Ctd)
Slab (Ctd)
Rfmt.

concrete slab — mesh reinforcement

Concrete cover for reinforcement

overall dims. 10·000 × 6·000
−conc. cover 2/0.050· 0·100   0·100
                    9·900   5·900

9·90
5·90

Fabric rfmt. for in situ conc. to BS. 4483 ref. A192 wgh. 3·02 kg/m² with 150 mm laps at all passings.

SECTION.
Reinforcement must not be exposed to the elements etc., otherwise it will rust and spall the concrete. Therefore an adjustment is made for the concrete cover when measuring reinforcement.

SMM E30.4.1.0.0

Beams
Main beam   Length.
                          6·800
−b.i. ends 2½/ 0·215· 0·215
                          5·785

Sec. beam   10·000
−b.i. ends a.b. : 0·215
− main beam : 0·200· 0·415
                          9·585.

N.B. Conc. in beams added back to slab R.C. Flat Rf. 2.

As the concrete to beams is poured at the same time as the concrete to slab, the volumes of concrete are added together and described as concrete in slabs. (See SMM E10.5–D4.) The thickness range is based on the slab thickness only. (See SMM E10.5–M2.)

R.C. Flat Rf. 5

Constr (Ctd)
Beams (Ctd)
Fmwk.

Main beams   girth
               soffit   0·200
  + sides   2/0·350: 0·700
                              0·900

Sec beams   total length
   fmwk. dim   9·570
    – main beam   0·200
                    9·370
                    girth
        soffit   0·150
  + sides 2/0·200· 0·400
                    0·550.

Formwork for
secondary beams

Formwork for
main beam.

Do not deduct main
beam formwork for
intersection – see note below.

DIAGRAM OF FORMWORK TO BEAMS.

5·57
0·90
9·37
0·55

Fmwk. for in situ. (main beam
conc. beams
attached to slabs (sec. beam
reg. shape, rectangular,
hgt. to soff. exc 3·00m
& n.e. 4·50m.
          (In Nr 4)

SMM E20.13.1.1.2

Where the secondary beam meets
the main beam no deduction of
formwork is made, but this
junction constitutes the start
of a new member. (See
SMM E20.13.1.1.2–M11.)

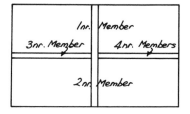

	1nr. Member	
3nr. Member		4nr. Members
	2nr. Member	

SOFFIT OF SLAB.

R.C. Flat Rf. 6.

Constr (Ctd)
Beam (Ctd)
Fmwk. adj. to slab.

5·57		Dolt fmwk. to (main beam
0·20		soff. of slab
9·37		a.b. (sec beam
0·15		

SMME 20.8.1.1.2

Reinforcement.

Length of bar rfcmt. in beams.
main beams   5·785
 – conc. cover   2/0·050:0·100
              5·685
+ hooks   2/9/0·016 · 0·288
              5·973

sec. beams   10·000
 – bwk.   2/0·215:0·215
– conc. cover   ab : 0·100:0·315
              9·685
+ hooks   2/9/0·012 = 0·216
              9·901

Usually the weight of reinforcement is obtained from bending schedules produced by the structural engineer and reinforcement is not often measured. Allowances must be made for hooks, bends, bobs, etc. when measuring reinforcement. The allowance for a semicircular hook on round bars is nine times the diameter of the bar. (See BS 4466.)

N.B. The secondary beam reinforcement goes through the main beam.

R.C. Flat Rf. 7.

Constr (Ctd)
Beams (Ctd)
Rfmt. (Ctd)

2/5.97

Rfmt. for in situ (main beam
conc. mild steel bars
to B.S. 4449, 16mm.dia.
str.
Wgt. up × 1.579 kg/m
= _____ kg.

SMM E30.1.1.1.0

All bar reinforcement is given
by weight in the bills of
quantities and therefore the
measurement must be weighted
up on dimensions or abstract.

sec. beams

2/9.90

Ditto, 12 mm dia. str.
Wgt. up × 0.888 kg/m
= _____ kg.

SMM E30.1.1.1.0

Upstands

Girth   10.000
                6.000
            2/16.000
ext. girth   32.000
-passings 4/2/£/0.150 = 0.600
& girth   31.400

31.40
0.15
0.25

In situ conc. (25N/20mm
agg.) upstand

SMM E10.14.0.0.0

Name 7   date   Name of Project.

R.C. Flat Rf. 8

Constr. (Ctd)
Upstand(Ctd)
Fmwk.

¢ girth		31.400
-passings 4/2/ⁿ/0.150		0.600
int. girth.		30.800

30.80	Fmwk. for in situ conc. sides of upstands plain vert. hgt. ≤ 250 mm.	SMM E20.4.1.2.0

NB.   Fmwk. to ext. face included with fmwk. to edge of slab. see Rf. 3.

Coverings

	10.000	×	6.000
-upstand ²/0.150: 0.300			0.300
	9.700		5.700

Asphalt is measured the area in contact with the base. (See SMM J21.1–M3.)

9.70 5.70	Mastic asp. rfg. 25mm thick to BS 988 limestone agg. in 2 cts. laid bkg. jt. with isol. membrane of sheathing felt to BS 7474 A (i) lapped 50mm at all passings laid loose on c&s screed, finished with white spar chippings bedded in bit. compound, width > 300 mm to falls & cross falls. external.	SMM J21.3.4.1.0

<u>R.C. Flat Rf. 9.</u>

<u>Cvgs. (Ctd)</u>

9·70	1000 g. Polythene sheet	SMM P10.1.1.1.0
5·70	vapour barrier with
	welded jts. plain areas
	horiz.

<u>Screeds</u>

<u>Av. thickness</u>

min. next outlet 'v'   = 0·040

max. at pt. 'x'. min = 0·040

+fall = $\dfrac{max\ dim\ (diag.)}{80}$

   = $\dfrac{5·800\ (scaled)}{80}$ = 0·073

max thickness   = 0·113

pts. U. W & X $\dfrac{3}{}$ 0·113 = 0·339

              4) :0·379

<u>Av. thickness : 0·095</u>

The expanded lightweight aggregate screed provides the fall of the roof, and it is necessary to work out the average thickness of the screed based on providing a fall of 1 in 80.

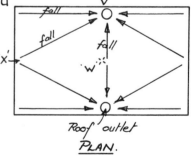

Roof outlet

<u>PLAN</u>.

As the roof is symmetrical the average thickness of the screed can be worked out over one quarter of roof, and applied to the whole screed.

R.C. Flat Rf. 10

<u>Cvgs (Ctd)</u>
<u>Screed (Ctd)</u>

9·70 5·70	Cmt. & lightweight agg. (1:5) as spec. insulating screed to roofs, to falls & crossfalls & to slopes ≤ 15° from horiz. av. 95mm thick in one ct. laid on poly. covered conc. <u>External</u>

SMM M10.6.1.1.0

All work is deemed to be
internal unless otherwise
described. (See SMM M10–D1.)

&

Cαs. (1:4) screed, rfs.
to falls & crossfalls & to
slopes ≤ 15° from horiz.
10mm thick in one ct.
laid on insulating screed.
<u>External</u>

SMM M10.6.1.1.0

<u>Vapour barrier at edge.</u>

<u>Girth</u>		9·700
		5·700
	²⁄	15·400
<u>int. girth of upstand</u>	=	30·800

<u>Height</u>	screed max.	0·113
	cαs topping	0·010
	top	0·075
		0·198

R.C. Flat Rf. 11

Cvgs. (Ctd)
Vapour Barrier (Ctd)

30·80		
0·20		

Poly. vapour barrier a.b.
plain areas . vert.

SMM P10.1.1.2.0

Kerbs.

Girths   Short sides

conc upstands   0·250
− screed max 0·113
  topping   0·010
  asp. ᵛrfg.   0·025 : 0·148
                           0·102
− ∠ fillet   0·050
                           0·052
+ ∠ fillet face : √0·050²×2 ∙ 0·071
                           0·123
+ top of upstand   0·150
                           0·273
+ passings ²/²/½/0·010 : 0·020
  girth on face   : 0·293

Long sides
min. girth as above : 0·293
max girth : min. 0·293
        + fall : 0·073 : 0·366
                2) 0·659
av. girth on face   = 0·330

The covering to kerbs along
the short side of roof will
have a regular girth for its
whole length, but the covering
to the kerbs to the two long
sides will have an average
girth because they follow the
fall of the roof.
The girth of coverings is the
girth on face. (See SMM J21–M4.)

Concrete
Upstand

girth on fac
Asphalt r[

SECTION.

<u>R.C. Flat Rf. 12.</u>

<u>Cvgs (Ctd)</u>
<u>Kerbs (Ctd)</u>

<u>Short sides</u>

<u>Length</u>		6·000
- to centre of kerb	2/½/ 0·150= 0·150	
		5·850

2/ 5·85

Mastic asp. a.b. cvgs. to
kerbs girth 225 - 300 mm
10mm thick in 2 cts. on conc.
trowelled smooth inc. ∠
fillet at base, arris &
wkg. into alum. trim.
<u>External</u>

SMM J21.11.3.0.0

<u>Long sides</u>

<u>Length</u>		10·000
- to centre of kerb a.b.		0·150
		9·850

2/ 9·85

Ditto girth > 300 mm av.
330 mm girth ditto
<u>External</u>

SMM J21.11.4.0.0

<u>Edge trim</u>

ext. girth of kerb a.b.		32·000
+passings	2/4/2/½/ 0·010=	0·080
<u>ext girth.</u>		32·080

Name of Project.

date

Name 12

<u>R.C. Flat Rf. 13.</u>

<u>Cvgs. (Ctd)</u>
<u>Edge trim (Ctd)</u>

32·08	Aluminium preformed trim 'Alutrim' type Z to suit asp. pas with alum. screws to conc. at 450 mm centres sleeved jts.	SMM J21.22.0.0.0

<u>Roof vents</u>

4	Bitumen compound rf. vents. fxd. in asp. rfg. inc. dishing screed	SMM J21.23.0.0.0

&

Asp. collar around pipes, standards & like members 75 mm dia. rf. vent. 125 mm high.   SMM J21.18.1.0.0

<u>NB.</u> One rf. vent is taken in each quarter of rf. This exceeds requirement of one for each 20 m².

Name of Project.

Name of

Name

date

Name 13

R.C. Flat Rf. 14.

Rainwater Goods.

2	Rainwater pipework ancillaries outlet 225 mm dia. with luting flange for asp. & screwed grtg. to BS 416 c.i. with outlet for 75 mm dia. pipe, bedding in preformed hole in conc. rf. in c.m (1:3) & caulked lead jt. to pipe.	SMM R10.6.2.1.1

&

| Fmwk. for in situ conc. holes girth 500 mm – 1·00 m, depth ≤ 250 mm conical circular shape 250 mm dia. at top & 100 mm dia at base. | SMM E20.27.2.1.3

Fabric reinforcement is deemed to include all cutting. (See SMM E30–C2.)

Screeds are deemed to include working into outlets, etc. (See SMM M10–C1(b).)

Asphalt is deemed to include working into outlet pipes etc. (See SMM J21–C1(c).) |

Name of Project

Name 14. date

R.C. Flat Rf. 15.

<u>Rainwater Gds (Ctd)</u>

<u>Pipes</u>

Flr.- clg. hgt.	3·600
offset.	0·150
into dr. collar	0·050
	3·800

2/<u>3·80</u> C.i. r.w.p. to BS. 416. str. 75 mm dia. with socketted caulked lead jts. fxd. with holderbatts p&s. to bwk at 2 m centres. SMM R10.1.1.1.1

2/<u>1</u> E.o. ditto fittings pipe > 65 mm dia. offset 150 mm projection. SMM R10.2.4.5.0

<u>To Take.</u>
① Decoration to pipe
② Joint to drain.

Name of Project. — *Name of Project.* date Name 15

13

Floors

Introduction

Within this section all types of suspended floors, e.g. timber upper floors, hollow ground floors, *in situ* pre-cast concrete floors and proprietary floors, would be included. However, in the worked example, single and double timber suspended floors are measured. Single floors are those where the floor joists simply span between supporting walls, whereas double floors are those which have a structural beam or beams to reduce the span of the joists.

With timber suspended floors, the floor finish, i.e. boarding or sheeting, is measured in this section, whereas with concrete suspended floors usually the screed and floor finishes are measured in the Internal Finishings section. The measurement of *in situ* concrete floor slabs is the same as that for the *in situ* concrete slab in the concrete flat roof example, Chapter 12.

Approach

Each building is divided into the following:

1. Ground floor – hollow ground floor, including sleeper walls, damp proof courses, air bricks, etc. Note: solid concrete ground floor slabs are usually measured in the Foundation section.
2. Upper floors – (a) single; (b) double, including structural beams.

It may be necessary to sub-divide timber floors according to rooms or groups of rooms.

Measure all floors over all openings, recesses, projections, etc., and then make adjustments. Any flooring in door openings is usually measured in the Doors section.

For measurement purposes floors are sub-divided into:

1. Construction, including any beams, sleeper walls, etc.
2. Coverings.
3. Adjustments for opening, etc.

Measurement

Steel beams

Steel Beams are given by weight in the bills of quantities and are measured in linear metres and then weighted up on the dimensions. Isolated structural beams must be so described, see SMM G12.5.1.1.

Steel beams are usually painted or treated at works and then again after erection to protect them from corrosion.

Timber

Structural timber is measured in linear metres and the form of construction required must be specified, otherwise it will be left to the discretion of the contractor, see SMM G20–S9 and Figures 13.1 and 13.2.

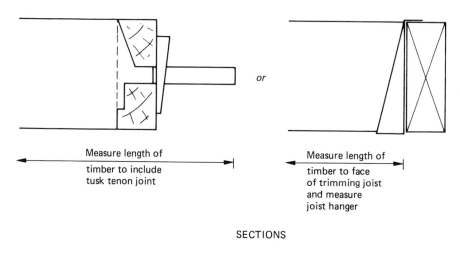

Measure length of
timber to include
tusk tenon joint

or

Measure length of
timber to face
of trimming joist
and measure
joist hanger

SECTIONS

Figure 13.1   *Sections*   Figure 13.2   *Sections*

Floor joists

Usually all joists in one building are the same depth to keep a level floor and ceiling throughout. The number of floor joists may be calculated as follows:

1. Calculate the distance at right angles to the span of the joists between the centres of the first and last floor joists. (Allow 25–50 mm for clearance between the face of the joist and the wall.)
2. Divide this distance by the designed centre of the floor joists to give the number of gaps between the joists.
3. Add one additional gap if there is any remainder.
4. Add one for the end floor joist to close the gap and give number of floor joists.

If there are any fixed points which affect the positioning of joists within a room, e.g. side of staircase opening, partition or chimney breast, etc., then the number of floor joists must be calculated between these points. Joists exceeding 6 m in length must be kept separate and so described.

Strutting

If the span of the joists exceeds 2.4 m a row of herringbone or solid strutting should be measured to help prevent the joists from twisting or warping. The strutting is measured in linear metres over floor joists and between wall faces.

Boarding

Floor boarding is measured in square metres and measured to the face of the structural walls. The method of jointing and method of fixing must be given in the description, as shown in Figures 13.3–13.5, otherwise it is left to the discretion of the contractor, see SMM K20–S2 and S12.

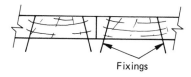

Figure 13.3   *Section – Square edge boarding*

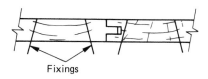

Figure 13.4   *Section – Tongued and grooved boarding*

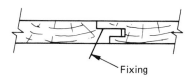

Figure 13.5   *Section – Rebated tongued and grooved boarding to allow for secret fixing*

If the floor is covered with a manufactured board flooring then it may be necessary to provide pieces of timber (noggins) in between the joists to support the edges of the board. It may be necessary to specify different quality manufactured boards in certain rooms, e.g. WBP boards in potential wet areas such as bathrooms.

Adjustment of openings

Adjustment of openings comprises:

1. Deduction of the overmeasure for trimmed joists (making due allowance for jointing or joist hangers).
2. Insertion of trimmer joists.
3. Replacement of a standard joist with a trimming joist. Trimmer and trimming joists are usually the same depth as the standard joists but are normally 25 mm wider.
4. Fixings, if required.
5. Deduction of floor boarding.
6. Protection of edge of flooring by measuring a timber nosing.

Flow chart for timber upper floors

SMM ref & unit	Measurement

Start

Steel beam

G.12.5.1.1.0 t.	Weight of structural steel beam, viz. length of beam built into brickwork both ends × weight of universal beam in kilograms per metre

Painting steel beam before fixing

M.60.5.1.1.3 m²	Surface area of steel universal beam, viz. length of beam as used above × girth of surface of universal beam

Painting steel beam after fixing

M.60.5.1.1.0 m²	Exposed surface area of steel universal beam, viz. same area as above less surface area of beams built into wall

Precast concrete padstones

F.31.1.1.0.0 nr.	Number of padstones

Softwood plates

G.20.8.0.1.0 m	Length of plates bolted to steel beam, viz. length measured between wall faces × two for both plates

Bolts

G.20.25.1.0.0 nr.	Number of bolts, viz. calculate number of bolts at 500 mm centres along web of steel beam

Softwood floor joists

G.20.6.0.1.0 m	Length of floor joists in all rooms (measured gross over recess, chimney breast and staircase well. Adjustments made later), viz. length of joists measured between inside face of walls plus allowance for building into internal walls and fixing with joist hangers to external walls × the number of joists at 375 mm centres with due allowance made for additional joists as required at fixed points, e.g. chimney breasts etc. Repeated for each room

Continued

SMM ref & unit	Measurement

Adjustment of softwood floor joists for recess

G.20.6.0.1.0 m	Deduct length of floor joists measured gross over recess, viz. length of recess plus allowance for building in end × number of joists at 375 mm centres

Steel joist hangers

G.20.21.1.0.0 nr.	Number of joist hangers, viz. number of end of joists next to external walls. Measured over chimney breast, adjustment made later

Herringbone strutting

G.20.10.1.1.0 m	Length of strutting in all rooms, viz. length measured between inside wall faces for each room

Fixing straps to external walls

To take note	Not measured in this example

Floor boarding

K.20.2.1.1.0 m²	Area of flooring in all rooms (measured gross over recess, chimney breast and staircase well. Adjustments made later.), viz. area of all rooms measured to inside face of walls

Adjustment of floor boarding for recess

K.20.2.1.1.0 m²	Deduct area of flooring measured over recess

Adjustment for chimney breast

G.20.6.0.1.0 m	Deduct length of trimmed floor joists measured over chimney breast, viz. length of chimney breast measured from room wall face to end of housed joint with trimmer joist × number of floor joists at 375 mm centres between trimming joists

Ditto

G.20.6.0.1.0 m	Deduct length of floor joists replaced by trimming joists, viz. same as length of floor joists measured in room × two for joists both sides of chimney breast

Continued

Flow chart for timber upper floors – continued

SMM ref & unit   Measurement

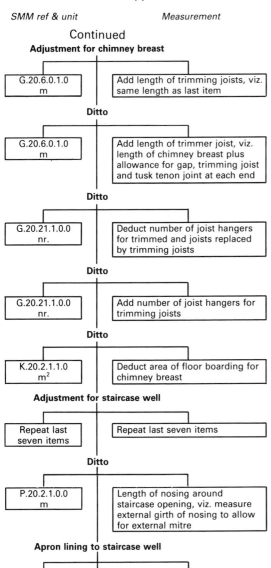

Continued

Adjustment for chimney breast

G.20.6.0.1.0
m
| Add length of trimming joists, viz. same length as last item

Ditto

G.20.6.0.1.0
m
| Add length of trimmer joist, viz. length of chimney breast plus allowance for gap, trimming joist and tusk tenon joint at each end

Ditto

G.20.21.1.0.0
nr.
| Deduct number of joist hangers for trimmed and joists replaced by trimming joists

Ditto

G.20.21.1.0.0
nr.
| Add number of joist hangers for trimming joists

Ditto

K.20.2.1.1.0
m²
| Deduct area of floor boarding for chimney breast

Adjustment for staircase well

Repeat last seven items
| Repeat last seven items

Ditto

P.20.2.1.0.0
m
| Length of nosing around staircase opening, viz. measure external girth of nosing to allow for external mitre

Apron lining to staircase well

To take note
| Not measured in this example

Check drawings and taking-off list for outstanding items

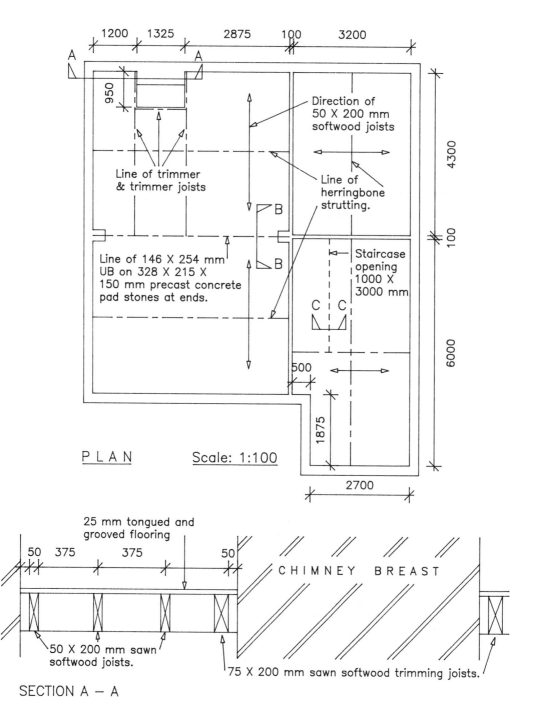

PLAN   Scale: 1:100

25 mm tongued and
grooved flooring

50 X 200 mm sawn
softwood joists.

75 X 200 mm sawn softwood trimming joists.

CHIMNEY   BREAST

SECTION A – A

TIMBER   UPPER   FLOOR

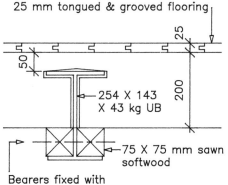

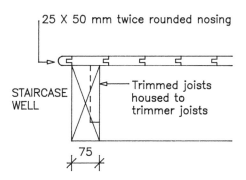

SECTION B – B   Scale: 1:10   SECTION C – C

SPECIFICATION

1. Universal beam to be BS4 Part 1 and painted with
   1 coat zinc chromate primer.
2. All sawn softwood to be treated with Tanalith.
3. Joists fixed to external walls with joist hangers
   manufactured by: Messrs'X. Those bearing on internal
   walls to be built in and trimmed joists housed to
   trimmer joists.
4. The herringbone strutting to be of two rows of
   25 X 50 mm sawn softwood.
5. Floor boarding to have tongued and grooved joints
   and each board twice nailed to each joist with
   2 Nr 10 gauge floor brads 50 mm long.

TIMBER UPPER FLOOR

Drawing Number.

Name of Project

date

Name 1

Timber Upper Floors 1

Taking Off List

Construction

    1. Beam   a. Universal Beam
                   b. Surface Treatments.
                   c. Padstones
                   d. Softwood Plates
                   e. Bolts.

    2. Joists   a. Joists
                   b. Hangers
                   c. Strutting

Coverings

    1. Boarding

Adjustments

    1. Chimney Breast   a. Construction.
                         b. Coverings.

    2. Staircase Well   a. Construction.
                         b. Coverings

Page Nrs.

~~1~~	~~11~~
~~2~~	~~12~~
~~3~~	13
~~4~~	14
~~5~~	15
~~6~~	16
~~7~~	17
~~8~~	18
~~9~~	19
~~10~~	20.

Timber Upper Flrs. 2.

Construction

Chimney breast.

PLAN.

Beam - Rm. 1.

<u>Length</u>

	1·200
	1·325
	2·875
	5·400

B.i. ends. part. 0·100

ext. wall <u>0·103 · 0·203</u>

5·603

5·60	Isolated structural steel member to B.S. 4 Pt. 1. plain U.B. member 254 × 143 mm beams. Wgt. up × 43 kg/m = _____ kg.

The floor in Room 1 is called a double floor, having a beam midway which splits the floor into two and reduces the span and therefore the depth of joists used.

SMM G12.5.1.1.0

Structural metal members are billed by weight. (See SMM units.) Measurement are converted to weight in the dimensions and reduced to tonnes in the bills of quantities.

Timber Upper Flrs. 3.

Constr. (Ctd)
Beam (Ctd)
Surf. treatments

girth of beam. flange 4/0·143·0·572
            – web   2/0·007·0·014
                          0·558
            + web   2/0·254·0·508
                          1·066

SECTION.

	5·60	Pp. wire brush & pt. 1 ct.	SMM M60.5.1.1.3
	1·07	of zinc chromate primer	

of zinc chromate primer
on structural metal work
gen. surf. girth > 300 mm
application on site to members
prior to fxg.

&

Pp & pt. 1 ct. ditto gen.
surf. girth > 300 mm.   SMM M60.5.1.1.0

Adj. for ends bi

2/0·10   Ddt ditto   (ends   SMM M60.5.1.1.0
  1·07
2/0·33   (on padstone   Adjustment for building in
  0·14   ends of universal beam.

Name 3   date   Name of Project.

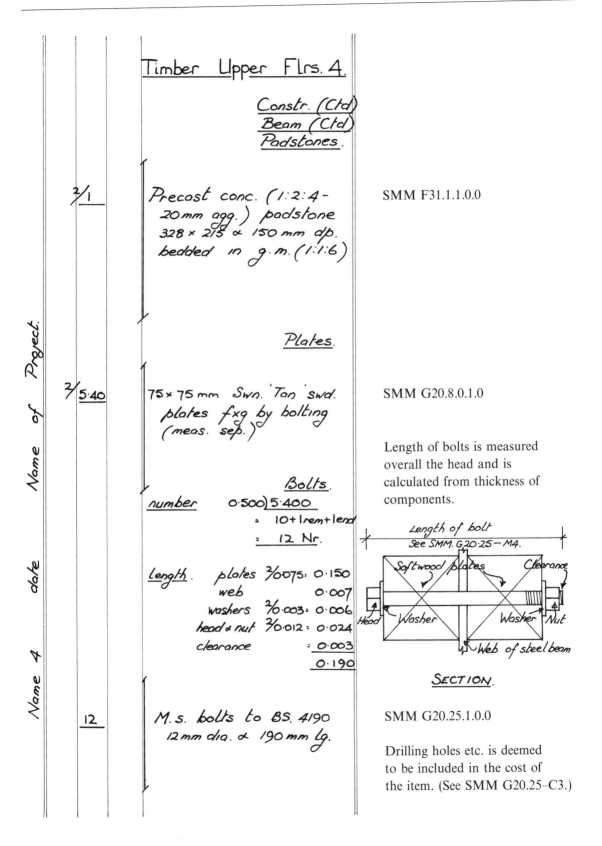

Name of Project.   Name of   date   Name 4

Timber Upper Flrs. 4.

<u>Constr. (Ctd)</u>
<u>Beam (Ctd)</u>
<u>Padstones</u>.

2/1

Precast conc. (1:2:4 -
20 mm agg.) padstone
328 × 215 × 150 mm dp.
bedded in g.m. (1:1:6)

SMM F31.1.1.0.0

<u>Plates</u>.

2/5·40

75 × 75 mm Swn. 'Tan' swd.
plates fxg by bolting
(meas. sep.)

SMM G20.8.0.1.0

Length of bolts is measured
overall the head and is
calculated from thickness of
components.

<u>Bolts</u>.

number   0·500)5·400
     = 10+1rem+1end
     = 12 Nr.

<u>Length</u>. plates 2/0·075: 0·150
     web   0·007
     washers 2/0·003: 0·006
     head & nut 2/0·012: 0·024
     clearance   : 0·003
       0·190

<u>SECTION</u>.

12

M.s. bolts to BS. 4190
12 mm dia. × 190 mm lg.

SMM G20.25.1.0.0

Drilling holes etc. is deemed
to be included in the cost of
the item. (See SMM G20.25–C3.)

<u>Timber Upper Flrs. 5</u>

<u>Constr. (Ctd)</u>
<u>Joists.</u>

<u>Rm 1</u> - above beam   <u>Number</u>
<u>Left hand side of breast</u>   1·200
    - gaps ²/0·050 = 0·100
    - ¢ jsts ½/0·050 = 0·025
    " " ½/0·075 = 0·038. <u>0·163</u>
      0·375) <u>1·037</u>
       = 2+1 rem+1 end
       = 4 Nr.
<u>Chimney breast</u>   1·325
    + gaps ²/0·050 = 0·100
    + jsts ²/0·075 = <u>0·075</u> = 0·175
      0·375) 1·500
       4 - 1 end.
       = 3 Nr.
<u>Right hand side of breast.</u>   2·875
    - gaps & jsts as l.h.side = <u>0·163</u>
      0·375) 2·712
       7+1 rem+1 end
       = 9 Nr.
       <u>Length</u>
       4·300
   + part. ²/0·100 = 0·050
       4·350
   - web ²/0·007 = <u>0·004</u>
       <u>4·346</u>

<u>Rm 1</u> - below beam   <u>Number</u>
      5·400
   - gaps ²/0·050 = 0·100
   - jsts ²/0·050 = 0·050 = 0·150
     0·375) 5·250
      14+ 1 end
      = 15 Nr.

Name 5   date   Name of Project.

The distance between the centre lines of the first and last joists is calculated in each room and then divided by the design centres of joists. One extra is added for any remainder and one is added for the end joist to give the total number of joists required in each room. Where certain joists have fixed positions, i.e. trimming joists next to openings, then the number of joists is calculated either side of these fixed points, as the total number can be affected.

Name of Project.   date   Name 6

Timber Upper Flrs. 6

Constr (Ctd)
Jsts. (Ctd)

Rm 1 - below beam (ctd)   Length.
| | 6·000 |

+ part.   ½/0·100 :   0·050
| | 6·050 |

− web ⅔/0·007 : 0·004
recess   : 1·875 : 1·879
| | 4·171 |

Rm 2.   Number
| | 4·300 |

− gaps & jsts as Rm 1. : 0·150
| 0·375 | 4·150 |
| : 11 + 1 rem + 1 end |
| : 13 Nr. |

                        Length
| | 3·200 |

+ part. b.i. end.   0·190
| | 3·300 |

Rm 3 -   Number
Staircase well   3·000

+ jst.   ½/0·075 ·   0·038
| | 3·038 |

− gap   0·050
− jst. ⅔/0·050· 0·025 · 0·075
| 0·375 | 2·963 |
| : 7 + 1 rem + 1 end |
| : 9 Nr. |

Remainder of rm. 3.   6·000
        − s/c well   3·000
| | 3·000 |

− gap & jst. ab. : 0·075
− trimmer ½/0·075· 0·038   0·113
| 0·375 | 2·887 |
| : 7 + 1 rem |
| : 8 Nr. |

Timber Upper Flrs 7.

Constr (Ctd)

Jsts (Ctd)

$\frac{4}{9.3}$/4·35	50 × 200 mm Swn.	(Rm. 1.		SMM G20.6.0.1.0
15/4·17	'Tan.' swd. flr.	(,, 1		
13/3·30	members.	(,, 2		The joists in Room 1 are in
$8.\frac{9}{}$/3·30		(,, 3		two lengths split by the beam

The joists in Room 1 are in two lengths split by the beam and do not exceed 6 m in one continuous length.

Rm.3- adj. for recess.

Number

0·375) 1·875

5 Nr.

Length

0·500

+ part. bi. end 0·100

0·600

5/0·60	Ddt 50 × 200 mm ditto.	SMM G20.6.0.1.0

Hangers.

$\frac{4}{9.3}$/1	2.5mm. Gal. m.s.	(Rm 1.	SMM G20.21.1.0.0
15/1	hangers manu.	(,, 1	
13/1	by Messrs 'X' to	(,, 2	
$8.\frac{9}{5}$/1	suit 50 × 200 m.m.	(,, 3	
5/1	jsts.	(,, 3	Five joists are fixed to
		(recess	external wall at both ends

Five joists are fixed to external wall at both ends within the recess.

			Timber Upper Flrs. 8.
			<u>Constr (Ctd)</u> <u>Strutting.</u>

<table>
<tr><td>³⁄<u>5·40</u>
<u>4·30</u>
<u>6·00</u></td><td>25 × 50mm Swn. Tan (Rm 1
swd. jst. strutting (" 2
herringbone to (" 3
200mm. dp. jsts.</td><td>SMM G20.10.1.1.0

Strutting should be taken where the span of the joists exceeds 2.40 m.</td></tr>
</table>

	<u>To Take</u> Metal straps between joists and flank walls.	The Building Regulations require metal straps between floor joists and the flank walls of buildings. If something is obviously missing from the drawings then an entry must be made in the query list and a 'To Take' note in the dimensions.

<u>Boarding.</u>	
Rm.l. 4·300	
+part 0·100	
6·000	
10·400	
− recess 1·875	
8·525	

<table>
<tr><td><u>5·40</u>
<u>8·53</u>
<u>3·20</u>
<u>4·30</u>
<u>3·20</u>
<u>6·00</u></td><td>25mm Wrot swd. timber(Rm 1.
bd. flrg. width ⧸300mm
t&g., each bd. twice (" 2
nailed to every swd.
jst. with 2 Nr. 10g. (" 3
flr. brads 50mm. lg.
with splayed heading joints
occurring over a jst.</td><td>SMM K20.2.1.1.0

The term *boarded* gives the contractor an idea of the width of boards used in the flooring. *Narrow strip* is the other term used (see SMM K21), but it may be better to clarify the width of boards required by stating the cross-sectional dimensions.</td></tr>
<tr><td><u>0·50</u>
<u>1·88</u></td><td>Ddt 25mm ditto (recess
Rm 3.</td><td>SMM K20.2.1.1.0</td></tr>
</table>

Timber Upper Flrs. 9.

Adjustments
Rm. 1. Chimney breast.

PREAMBLE   NOTE
Housed joints are required
between trimmed & trimmer
joists & tusk tenons between
trimmer & trimming joists.

The method of jointing must be
stated, otherwise it is left to
the discretion of the
contractor. (See SMM G20.S9.) If
traditional jointing is required,
sufficient timber must be
measured as in this example.

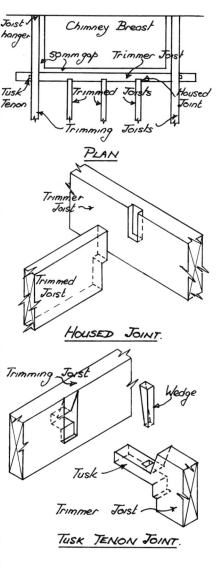

Trimmed Joists.

Length - chimney breast   0·950
+ gap   0·050
+ trimmer jst. 0·075
- housing   0·025 = 0·050·0·100
                            1·050

number   as calc. before = 3/Nr.

Trimmer Joist.

Length- chimney breast 1·325
+ gaps   2/0·050:0·100
+ trimming jsts. 2/0·075 · 0·150
+ horns   2/0·150· 0·300
                         1·875

PLAN

HOUSED JOINT.

TUSK TENON JOINT.

Timber Upper Flrs. 10.

Adjustments (Ctd)
Chimney breast (Ctd)

3/1·05	Ddt 50 × 200 mm (trimmed jsts	SMM G20.6.0.1.0
3/4·35	swd. flr. members (trimming	
	a.b. (jsts. Rm l.	
8/1·15	(trimmed jsts Rm 3	
3·30	(trimming jsts " 3	

Trimmer & Trimming Joists.

2/4·35	Add 75 × 200 mm (Rm 1.	SMM G20.6.0.1.0
1·88	ditto. (" 1	
3·30	(Rm 3	
3·33	(" 3.	

Jst. hangers.

2·3/1	Ddt jst. hangers (Rm 1	SMM G20.21.1.0.0
1	a.b. to suit 50 × (" 3	
	200 mm jsts.	

2/1	Add ditto to suit (Rm 1.	SMM G20.21.1.0.0
1	75 × 200 mm. jsts. (" 3	

Name of Project. date Name 10

Timber Upper Flrs. 11

Adjustments (Ctd)
Chimney breast (Ctd)

1·33 0·95 1·00 3·00	*Ddt* 25 mm wrot (Rm 1. swd. flrg a.b. (Rm 3.

SMM K20.2.1.1.0

Staircase Well.

Trimmed Joists
Length - staircase well   1·000
+ part. b.i. end   0·100
+ trimmer jst.   0·075
- housing   0·025 = 0·050
                   1·150

number - as calc before = 8/Nr.

NOTE
Adj. for s/c well are added
back to Timber Upper Flrs
10 except nosing.

Trimmer Joist
Length - s/c well   3·000
+ part. b.i. end   0·100
+ trimming jst.   0·075
+ horn   0·150
               3·325

Name   Name of Project.   11   date

Adding back dimensions saves time in both taking-off and working up as it collates like items together. Usually it is done in a different coloured ink so that the dimensions can be checked.

Timber Upper Flrs. 12.

<u>Adjustments (Ctd)</u>
<u>Staircase Well (Ctd)</u>

<u>Nosing.</u>

girth

	1·000
	3·000
	4·000

tongue 2/6·010·0·020

4·020

| 4·02 | 25 × 50 mm Wrot swd nosing twice rdd. , twice reb. & tongued to edge of flr. inc. gr. in flr. | SMM P20.2.1.0.0 |

SMM P20.2.1.0.0

Fixing must be stated. (See SMM P20.2–S8.)

<u>To Take</u>
Apron lining to s/c well.

Name of Project.

Name date 12

Name

Appendix 1   List of abbreviations

ab	as before	chfd	chamfered
abd	as before described	chmny	chimney
ABS	acrylonitrile butadiene styrene	ci	cast iron
add	additional	circ	circular
adj	adjacent	cist	cistern
agg	aggregate	cl or ₵	centre line
AI	Architect's Instruction	clng	ceiling
arnd	around	clng jst	ceiling joist
art	artificial	cm	cement mortar
asb	asbestos	coe	curved on elevation
asph	asphalt	col	column
av	average	conc	concrete
awp	as work proceeds	cp	chromium plated
		cpd	cupboard
bal	baluster	cpvc	chlorinated polyvinyl chloride
b & p	bed and point	crse	course
bdd	bedded	csk	countersunk
be	both edges	csmnt	casement
bf	before fixing	ct	cement
bi	built in	ct & b	cut tooth and bond
bkt	bracket		
bldg	building	d/d	delivered
blk	block	ddt	deduct
blwk	blockwork	dh	double hung
bm	birdsmouth	dia	diameter
bma	bronze metal antique	dist	distance
bn	bull nosed		distemper
boe	brick on edge	dp	deep
brd	board	dpc	damp proof course
brdg	boarding	dpm	damp proof membrane
brk	brick	drwg	drawing
brs	bearers	ds	double seal
bs	both sides		
BS	British Standard	ea	each
BSC	British Standard Channel	egl	existing ground level
bsmt	basement	eml	expanded metal lathing
BSUB	British Standard Universal Beam	EO	extra over
		ES	earthwork support
bwk	brickwork	exc	excavate
		excvn	excavation
ca	cart away	ext	external
c & p	cut and pin	extg	existing
c & s	cups and screws	f & b	framed and braced
cb	common bricks	FAI	fresh air inlet
cc	centres	fc	fair cutting
ccn	close copper nailing	fcngs	facings
ce	cleaning eye	fdns	foundations

ffl	finished floor level	l & b	ledged and braced
fin	finished	lab	labour
fl	floor level	l & c	level and compact
	florination level	lp	large pipe
fl & b	framed ledged and braced		
fr	frame	mat	material
frd	framed	mg	make good
ftd	fitted	mh	manhole
fwk	formwork	mort	mortice
		ms	measured separately
galv	galvanized		mild steel
gen surf	general surfaces	msrd	measured
gi	galvanized iron	mupvc	modified unplasticised polyvinyl
gf	ground floor		chloride
gl	ground level	ne	not exceeding
gm	gauged mortar	nsg	nosing
	gunmetal	nts	not to scale
gms	galvanized mild steel		
grano	granolithic	ø	diameter
grth	girth	o/a	overall
gwg	georgian wire glass	ocn	open copper nailing
gyp	gypsum	opg	opening
		org	original
hbs	herringbone strutting	os	one side
hbw	half brick wall	③	three oils
hc	hardcore		
hdg jnt	heading joint	p & s	plugged and screwed
hm	hand made		plank and strut
hn & w	head nut and washer	pbrd	plasterboard
horiz	horizontal	PC	Prime Cost
HP	High Pressure	plas	plaster
hr	half round	plstrd	plastered
ht	height	pm	purpose made
	hollowtile	po	prime only
hw	hardwood		planted on
	hollow wall	pol	polished
		pr	pair
		prep	prepare
IC	Inspection Chamber	proj	projection
inc	including	prov	provisional
int	internal	pt	point
inv	invert	ptd	pointed
		ptg	pointing
jap	japanned	ptn	partition
jst	joist	PVA	Polyvinyl acetate
jt	joint	PVC	Polyvinyl chloride
jtd	jointed	pvg	paving
kps	knot, prime and stop	qt	quarry tile

r & s	render and set		T	tee
rad	radius		t & g	tongued and grooved
rc	reinforced concrete		t & r	treads and risers
rdd	rounded		tc	terra cotta
re	rodding eye		temp	temporary
reb	rebated			
reinf	reinforced		UB	Universal Beam
retd	returned		UC	undercoat
rl	reduced level			Universal Column
rljt	red lead joint		uPVC	unplasticized polyvinyl chloride
rme	returned mitred end			
ro	rough		vent	ventilation
roj	rake out joints		vert	vertical
rsc	rolled steel channel		vit	vitrified
rsj	rolled steel joist			vitreous
rwp	rainwater pipe		VO	variation order
			vp	vent pipe
saa	satin anodised aluminium			
s & f	supply and fix		w & p	wedge and pin
s & l	spread and level		WBP	water and boil proof
sbj	soldered branch joint		wdw	window
SBR	styrene butadiene rubber		wg	white glazed
sc	stop cock		wi	wrought iron
sd	screw down		wp	waste pipe
se	stopped end		ws	written short
segtl	segmental		wt	weight
sg	salt glazed		wthd	weathered
sjt	soldered joint			
sk	sunk		× falls	crossfalls
sktg	skirting		× grain	cross grain
sl	short length		× tgd	cross tongued
soff	soffit		× tng	existing
sp	small pipe			
spec	specification		mm	millimetre
sq	square		m	metre
ss	stainless steel		m²	square metre
stg	starting		m³	cubic metre
susp	suspended		No or Nr	number
svp	soil vent pipe		kg	kilogramme
sw	softwood		t	tonne
	stoneware			
swn	sawn			

Appendix 2   Query list

No. _____   Job Name: _____

Date: _____   Attn. of: _____

Query (including Drawing Ref)	Answer (including Drawing Ref)	Answer By

Signed Q S   Signed Architect/Engineer

3 Copies – 2 to Arch/Eng (1 for return)   1 for Q S File

Appendix 3   Reinforcement constants

	Weight
Bar diameter	kg/m
6 mm	0.222
8 mm	0.395
10 mm	0.616
12 mm	0.888
16 mm	1.579
20 mm	2.466
25 mm	3.854
32 mm	6.313
40 mm	9.864
Fabric	kg/m²
Ref A98	1.54
Ref A142	2.22
Ref A193	3.02
Ref A252	3.95
Ref A393	6.16
Ref B196	3.05
Ref B283	3.73
Ref B385	4.53
Ref B503	5.93
Ref B785	8.14
Ref B1131	10.90
Ref C283	2.61
Ref C385	3.41
Ref C503	4.34
Ref C636	5.55
Ref C785	6.72

Appendix 4 Bending schedule

MEMBER	BAR MARK	TYPE & SIZE	No. OF MBRS.	No. IN EACH	TOTAL No	LENGTH OF EACH BAR+ mm.	SHAPE CODE	BENDING DIMENSIONS				
								A* mm	B* mm	C* mm	D* mm	E/r* mm
FOUNDATIONS												
	01	T16	1	46	46	2000	37	1800				
	02	T10	1	159	159	1350	60	190	380			
	03	T16	1	8	8	2300	37	2100				
	04	T16	1	6	6	3700	65	3700				6720
	05	T12	1	298	298	1800	38	200	1445			
PILE CAPS Nos 1,2,3,4,5 & 6												
	06	T16	6	10	60	1625	38	400	900			
300 × 200 LINTELS												
	07	T10	2	6	12	1875	35	1675				
	08	R8	2	11	22	700	60	95	170			

JOB No. 123   **BAR SCHEDULE REFERENCE**   **DRAWING No.** 001   **SCHED No.** 01   **REV**

CONTRACTOR WE BUILD IT LTD.

SITE NEWTOWN

HAVECALC & WILL TRAVEL
CONSULTING ENGINEERS

ALL BENDING DIMENSIONS IN ACCORDANCE WITH B.S. 4466

*SPECIFIED TO NEAREST 5mm
†SPECIFIED TO NEAREST 25mm

BY   **CHECKED**

Appendix 5 Purpose-made bricks

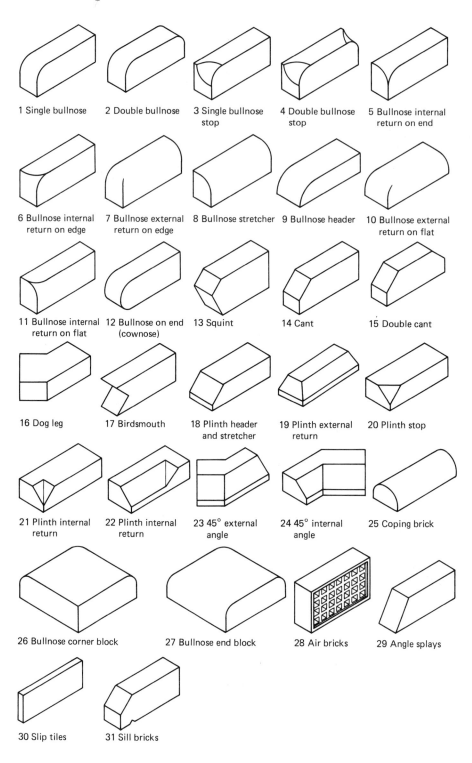

1 Single bullnose

2 Double bullnose

3 Single bullnose stop

4 Double bullnose stop

5 Bullnose internal return on end

6 Bullnose internal return on edge

7 Bullnose external return on edge

8 Bullnose stretcher

9 Bullnose header

10 Bullnose external return on flat

11 Bullnose internal return on flat

12 Bullnose on end (cownose)

13 Squint

14 Cant

15 Double cant

16 Dog leg

17 Birdsmouth

18 Plinth header and stretcher

19 Plinth external return

20 Plinth stop

21 Plinth internal return

22 Plinth internal return

23 45° external angle

24 45° internal angle

25 Coping brick

26 Bullnose corner block

27 Bullnose end block

28 Air bricks

29 Angle splays

30 Slip tiles

31 Sill bricks

Appendix 6 Brick facework details

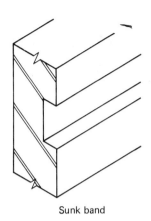

Sunk band

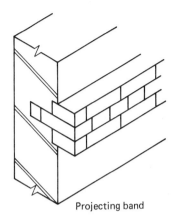

Projecting band

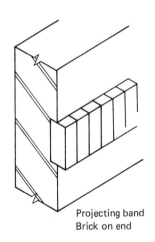

Projecting band
Brick on end

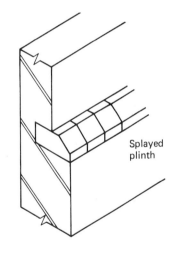

Splayed
plinth

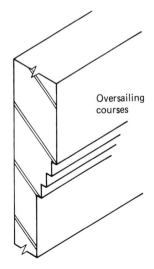

Oversailing
courses

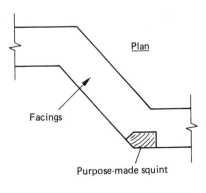

Plan

Facings

Purpose-made squint

Squint Angles

Appendix 7   Mathematical formulae

Appendix 7.

Formulae 1

Including entering in the dimension column.

2·00 5·00	Eg. 'A' two sets of measurements to build up the required unit – m².
2·00 5·00 3·00	Eg. 'B' three sets of measurements to build up the required unit – m³.

Linear.

2/π/5·00	Circumference of a circle (let r = 5·00m)
½/2/π/5·00	Circumference of a semi circle.

The units stated in the SMM for measured items should be represented correctly in the dimension column to allow for easy visual checking during the working-up process, e.g. 'A', an item measured in m², should have two sets of measurements written in the dimension column.

E.g. 'B', an item measured in m³, should have three sets of measurements written in the dimension column.

Or an instruction can be given in the description column to super or cube up linear or super dimensions.

N.B. One measurement written in the dimension column and the timesing column is used for the application of the formulae.

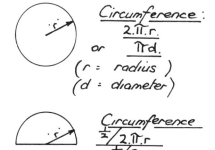

Circumference:
$$\frac{2.\pi.r}{}$$
or $\pi d.$
(r = radius)
(d = diameter)

Circumference
$$\frac{1}{2}/2.\pi.r$$
or $\frac{1}{2}/\pi d$

Appendix 7.

Formulae 2.

Linear (Ctd)

Length of arc.

$\frac{\theta}{360}/2\pi r \quad 5·00$

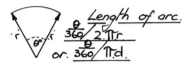

$\frac{Length\ of\ arc.}{\frac{\theta}{360}/2\pi r}$
or. $\frac{\theta}{360}/\pi d.$

Superficial.

N.B. Two measurements written in the dimension column.

Area of triangles

$\frac{1}{2}/10·00$
$\quad 5·00$

Area of triangle
(let b = 10·00m &
h = 5·00m)

(i)

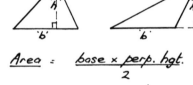

Area = $\frac{base \times perp.\ hgt.}{2}$

(b = base)
(h = perp. height)

(ii) Triangle with three sides
dimensioned.

(Let a = 5·00m, b = 4·00m &
c = 2·00m)

$$
\begin{array}{r}
& S \\
a = & 5·000 \\
b = & 4·000 \\
c = & 2·000 \\
\hline
2\,) & 11·000 \\
\hline
S = & 5·500 \\
\end{array}
$$

Area = $\frac{\sqrt{S(S-a)(S-b)(S-c)}}{}$
(S = half sum of sides)
(a, b & c = lengths of sides)

S: 5·500 : 5·500 : 5·500
−a: 5·000 −b: 4·000 −c: 2·000
S−a: 0·500 S−b: 1·500. S−c: 3·500

Area = $\sqrt{5·500. \ 0·500. \ 1·500. \ 3·500}$
= $\sqrt{14·438}$ = 3·800 m²

Sometimes it is easier to work out complicated formulae in waste, but the answer should be written in the dimension column showing the correct units by including × 1.00 m etc.

3·80
1·00

Area of triangle (3 sides given)

Formulae 3

Supers. (Ctd)

$\frac{1}{2}$/
Sin θ/ 5·00
        6·00

Area of triangle (2 sides
and include angle)
(let a = 5·00m, b = 6·00m
& ∠C = θ°)

a =   6·000
b =   9·000
    2) 15·000
        7·500

7·50
3·00

Area of trapezium (let
a = 6·00m, b = 9·00m and
h = 3·00m)

Area of triangles (Ctd)

(iii) *Triangle with two sides and included angle.*

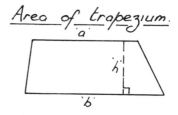

Area : $\dfrac{a.b.\sin\angle C}{2}$

Area of trapezium.

Area : $\dfrac{a+b}{2} \times h.$

(a & b = Lengths of parallel
sides & h = perp.
distance between them)

Area of Polygons.

N.B. Usually other shapes can
divided up into triangles and
the areas added together to
give the total area.
Irregular boundary lines are
straightened by drawing in
give and take lines over the
irregular line.

Appendix 7.

Formulae 4

Appendix 7.

Supers. (Ctd)

Area of Polygons (Ctd)

Eg. regular pentagon. (5 sides)

Area $\triangle AOB = \frac{1}{2} \times S \times h$

but $h = \tan 54° \times \frac{S}{2}$

$\qquad = 1.376 \,(\text{nat. tan}) \times \frac{S}{2}$

$\therefore$ Area $= \frac{1}{2} \times S \times 1.376 \times \frac{S}{2}$

$\therefore$ Area of regular polygon

$\qquad = 5 \times \frac{1}{2} \times S \times 1.376 \times \frac{S}{2}$

Area $= S \times S \times 1.720$

Area of regular pentagon.
(let S = 9.00m)

1.720 / 9.00 / 9.00

(6 sides) Area regular hexagon $= S \times S \times 2.598$
(7 sides)   "   "   heptagon $= S \times S \times 3.634$
(8 sides)   "   "   octagon $= S \times S \times 4.828$
(9 sides)   "   "   nonagon $= S \times S \times 6.182$
(10 sides)   "   "   decagon $= S \times S \times 7.694$

(S = length of sides)

Area of circle
(let r = 5.00m)

π / 5.00 / 5.00

Area of circle.
Same drawing as before
Area $= \pi r^2$

Area of semi circle
(let r = 5.00m)

$\frac{1}{2}\pi$ / 5.00 / 5.00

Area of semi-circle
same drawing as before
Area $= \frac{1}{2} \times \pi r^2$

<u>Formulae   5</u>

<u>Supers (Ctd)</u>

Area of sector of circle
(let 'r' = 5·00m)

$\frac{\theta}{360}/\pi$ / 5·00 / 5·00

Area of segment
(let 'r' = 5·00m)   (sector

$\frac{\theta}{360}/\pi$ / 5·00 / 5·00

<u>Ddt</u>   (triangle

$\frac{1}{2}/\sin\theta$ / 5·00 / 5·00

Area of bellmouth
(let 'r' = 5·00m)

$\frac{3}{14}$ / 5·00 / 5·00

<u>Appendix 7.</u>

<u>Area of sector.</u>
Same drawing as before
Area $= \frac{\theta}{360}/\pi r^2$

<u>Area of segment.</u>

A figure: triangle AOB with angle $\theta°$ at O, radii r.

Area of segment
= Area of sector
   − Area of △AOB.

Area $= \frac{\theta}{360}/\pi r^2 - \frac{1}{2}.a.b.\sin C$

<u>Area of bellmouth</u>

Area of bellmouth
= Area of square
   − Area of sector

Area $= r^2 - \frac{90}{360}/\pi r^2$

$= r^2 - \frac{1}{4} \times \frac{22}{7} \times r^2$

$= r^2 - \frac{11}{14} r^2$

$= \frac{3}{14} r^2$

<u>Ordinate Rules</u>

Irregular areas are divided
into a series of strips by
equally spaced ordinates which
are then measured and then one
of the following formulae
applied.

Appendix 7.

Formulae 6.

Supers. (Ctd)

Ordinate Rules (Ctd)

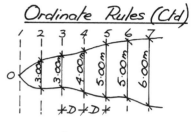

PLAN.

1st ord.	$\frac{0.000}{2}$	: 0.000
2nd ‥		: 3.000
3rd ‥		: 3.000
4th ‥		: 4.000
5th ‥		: 5.000
6th ‥		: 5.000
last ‥	$\frac{6.000}{2}$	: 3.000
		23.000

2·00
23·00

Area of irregular shape using trapezoidal rule (let D = 2.00m)

(i) **Trapezoidal rule**
(Assumes straight line boundary lines between ordinates)
(Any number of ordinates)

$$\text{Area} = D\left(\frac{1st + last\ ords.}{2} + A\right)$$

(D = common dist. bet. ords)
(A = sum of intermediate ords)

Ord. nr.	1st & last	Even nr.	Other odd nr.
1st	0.000	-	-
2	-	3.000	-
3	-	-	3.000
4	-	4.000	-
5	-	-	5.000
6	-	5.000	-
last	6.000	-	-
	6.000	12.000	8.000
		× 4 = 48.000	× 2 = 16.000
			48.000
			6.000
			70.000

(ii) **Simpson's rule.**
(Assumes slight curved to boundary lines between ordinates)
(Must have an odd number of ordinates and 1st ordinate is numbered one)

$$\text{Area} = \frac{D}{3}(1st + last\ ord. + 4 \times B + 2 \times C)$$

(D = common dist. bet. ords)
(B = Sum of even numbered ords)
(C = ‥ ‥ other odd ‥ ‥)

Appendix 7.

Formulae 7

Supers (Ctd)

$\frac{1}{3}$/2·00
7̲0̲·̲0̲0̲

Area of irregular shape using Simpson's rule.
(let D = 2·00m)

Volumes

N.B. Three measurements written in the dimension column.

$\frac{1}{3}$/5·00
4·00
6̲·̲0̲0̲

Volume of rectangular (a
base pyramid (b
(let a = 5·00m, b = (h
4·00m & h = 6·00m)

<u>Volume of pyramids or cones</u>

$Vol. = \frac{1}{3} \times$ area of base × perp-hgt.

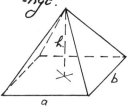

(a = width of base)
(b = length of base)
(h = perp. height)

$\frac{1}{3}$/$\frac{1}{2}$/6·00
7·00
6̲·̲0̲0̲

Volume of triangular (a
base pyramid. (b
(let a = 6·00m, b = 7·00m (h
& h = 6·00m)

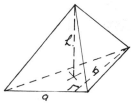

(a = base)
(b = perp. hgt. of base)
(ℓ = " " " pyramid)

Appendix 7.

Formulae 8

<u>Volumes (Ctd)</u>   <u>Volume of cones (Ctd)</u>

$\frac{1}{3}/\pi$

6·50
6·50
6·00

Volume of circular base (r
pyramid or cone.   (r
(let r = 6·50m &   (h
h = 6·00m)

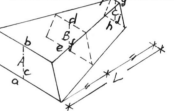

(r = radius of base)
(h = perp. height)

<u>1st Section</u>   <u>Mid Section</u>   <u>Last Section</u>

a = 6·000   d = 3·500   g = 1·000
b = 8·000   e = 5·000   h = 2·000
2)14·000   2)8·500   2)3·000
7·000   4·250   1·500

<u>Prismoidal formulae.</u>

$\frac{1}{6}/$ 30·00
7·00
3·00
$\frac{1}{6}/4/$ 30·00
4·25
2·50
$\frac{1}{6}/$ 30·00
1·50
2·00

Volume using
prismoidal formulae. (A
(let a, b, d, e, g & h
be as above, c =
3·00m, f = 2·50m,
j = 2·00m &
L = 30·00m)

(L
(A
(c
(L
(B
(f
(L
(c

(j

$Vol = \frac{L}{6}\left(A + 4 \times B + C\right)$
(A = Area of one end section)
(B =   "   " mid section)
(C =   "   " other end section)

OR.

7·00
3·00
4/ 4·25
2·50
1·50
2·00

Volume using prismoidal(A
formulae
Cube × $\frac{1}{6}$/30·00 =   m³ (B

(C

This method saves time in
writing dimensions and working
up.

Formulae 9

Volume (Ctd)

Ordinate Rules.

Trapezoidal and Simpsons rules can be used to calculate volumes of irregular shaped items. Use as for area measurement above, but the ordinates are measured areas instead of linear dimensions.

See ordinate drawing used above for calculation of areas.

Let the ordinates have the following calculated areas: —

Ord. nr.	Area m^2
1	0
2	5·000
3	3·000
4	6·000
5	5·000
6	7·000
7	9·000

Ord nr.		Area m^2	
1	$\frac{0}{2}$	:	0·000
2		:	5·000
3		:	3·000
4		·	6·000
5		:	5·000
6		:	7·000
7	$\frac{9·000}{2}$	:	4·500
			30·500

(i) *Trapezoidal rule*

$$Vol = D\left(\frac{1st + last\ ord.\ area}{2} + A\right)$$

$(D$ = common dist. bet. ords.$)$

$(A$ = sum of areas of intermediate ords$)$

2·00	
30·50	
1·00	

Volume of irregular item using trapezoidale rule
(let D = 2·000m)

The 1.00 m is included in the dimension to show that a cube unit is measured.

Formulae 10

Volume (Ctd)   Ordinate Rules (Ctd)

(ii) Simpson's Rule.

$$Vol = \frac{D}{3}(\text{1st+ last ord. area} + 4\times B + 2\times C)$$

(D : common dist. bet. ords)

(B : sum of areas of even numbered ords)

(C : sum of areas of other odd numbered ords)

Ord nr.	1st+last m^2	Even nr. m^2	Other odd nrs m^2
1st	0·000	-	-
2	-	5·000	-
3	-	-	3·000
4	-	6·000	-
5	-	-	5·000
6	-	7·000	-
7 (last)	9·000	-	-
	9·000	18·000	8·000
		×4=72·000	×2=16·000
			9·000
			72·000
			97·000

$$\frac{1}{3} \Big| \begin{array}{l} 2·00 \\ 97·00 \\ 1·00 \end{array}$$

Volume of irregular item using Simpson's rule   ①

(let D = 2·00m)

These are a selection of the more useful formulae. There are many more formulae which the student could add here, but most areas etc. can be built up using the basic formulae above.

Index